2015 IEEE 15th Topical Meeting on Silicon Monolithic Integrated Circuits in RF Systems

(SiRF 2015)

San Diego, California, USA
26-28 January 2015

IEEE Catalog Number:	CFP15SMI-POD
ISBN:	978-1-4799-8198-4

Copyright © 2015 by the Institute of Electrical and Electronic Engineers, Inc
All Rights Reserved

Copyright and Reprint Permissions: Abstracting is permitted with credit to the source. Libraries are permitted to photocopy beyond the limit of U.S. copyright law for private use of patrons those articles in this volume that carry a code at the bottom of the first page, provided the per-copy fee indicated in the code is paid through Copyright Clearance Center, 222 Rosewood Drive, Danvers, MA 01923.

For other copying, reprint or republication permission, write to IEEE Copyrights Manager, IEEE Service Center, 445 Hoes Lane, Piscataway, NJ 08854. All rights reserved.

***This publication is a representation of what appears in the IEEE Digital Libraries. Some format issues inherent in the e-media version may also appear in this print version.**

IEEE Catalog Number: CFP15SMI-POD
ISBN 13:   978-1-4799-8198-4

Additional Copies of This Publication Are Available From:

Curran Associates, Inc
57 Morehouse Lane
Red Hook, NY 12571 USA
Phone: (845) 758-0400
Fax: (845) 758-2633
E-mail: curran@proceedings.com
Web: www.proceedings.com

2015 IEEE 15th Topical Meeting on Silicon Monolithic Integrated Circuits in RF Systems (SiRF 2015)

San Diego, California, USA
26-28 January 2015

IEEE Catalog Number: CFP15SMI-POD
ISBN: 978-1-47998-198-4

SiRF 2015 Session List

Session List

- ❖ MO1B : SiRF Circuits and Applications - 1
- ❖ MO2B : mm-Wave and Higher Frequency Applications
- ❖ MO3B : RWS-SiRF Joint Session: Analysis and Arrays
- ❖ MO4B : RFSOI Technology and Applications
- ❖ TU1C : Topics in RF Modeling and Characterization Techniques
- ❖ TU3C : Power Amplifer Applications
- ❖ TU5A : RWS-SiRF Joint Session: RF & Internet of Things
- ❖ WE1C : Tunable and Reconfigurable Technologies
- ❖ WE2C : SiRF Circuits and Applications - 2
- ❖ WE3P : Joint RWW Interactive Poster Session: Transceivers and Front-end Technologies SOC and SiP

SiRF 2015 Table of Contents

MO1B: SiRF Circuits and Applications - 1

Chair: Larry Larson, Brown University — Co-Chair: Rahul Kodkani, QualComm
Venue: Grand Salon A, 08:00 – 09:40, Monday 26 January 2015

PAGE 1
MO1B-1
SIMMWIC Integration of Millimeter-Wave Antenna with Two Terminal Devices for Medical Applications *(Invited Paper)*
(Erich Kasper, Wogong Zhang)

PAGE 4
MO1B-2
A SiGe Differential 50ps Gaussian Pulse Generator for Sub-Sampling TDR Measurements
(Gregor Hasenaecker, Hans-Martin Rein, Klaus Aufinger, Nils Pohl, Thomas Musch)

PAGE 7
MO1B-3
A 6.5mW, Wide Band Dual-Path LC VCO Design with Mode Switching Technique in 130nm CMOS
(Jun Li, Ni Xu, Yuanfeng Sun, Woogeun Rhee, Zhihua Wang)

PAGE 11
MO1B-4
Design of Fully Integrated Receiver Front-End for VSAT Applications
(Ping-Yi Wang, Yun-Chun Shen, Min-Chih Chou, Yin-Cheng Chang, Te-Lin Wu, Da-Chiang Chang, Shawn S. H. Hsu)

SiRF 2015 Table of Contents

MO2B: mm-Wave and Higher Frequency Applications

Chair: Herman Schumacher, Ulm University — Co-Chair: Austin Chen, Skyworks Solutions
Venue: Grand Salon A, 10:10 – 11:50, Monday 26 January 2015

PAGE 14
MO2B-1
Review of Silicon-Based Millimeter-Wave Radio Frequency Integrated Circuits *(Invited Paper)*
(Huei Wang)

PAGE 15
MO2B-2
A 122–150GHz LNA with 30dB Gain and 6.2dB Noise Figure in SiGe BiCMOS Technology
(Roee Ben Yishay, Evgeny Shumaker, Danny Elad)

PAGE 18
MO2B-3
120GHz Low Power, High Gain, Wideband Active Balun for Chip-to-Chip Communication
(Chae Jun Lee, Hae Jin Lee, Dongmin Kang, In Sang Song, Hong Yi Kim, Seong Jun Cho, Joong Geun Lee, Inn-Yeal Oh, Chul Soon Park)

PAGE 21
MO2B-4
Electronic THz Transmissive Imaging System
(Wei-Cheng Chen, Chih-Wei Lai, Tzu-Chao Yan, Chun-Hsing Li, Tzu-Yuan Chao, Chien-Nan Kuo)

MO3B: RWS-SiRF Joint Session: Analysis and Arrays

Chair: Jeremy Muldavin, MIT Lincoln Laboratory — Co-Chair: Tommy Ellis
Venue: Grand Salon A, 13:30 – 15:10, Monday 26 January 2015

PAGE 24
MO3B-3
Intermittently Operating RF Frontend with 5ns Startup Time for 10Gbps Proximity Wireless Communication
(Naoki Kitazawa, Kaoru Kohira, Hiroki Ishikuro)

SiRF 2015 Table of Contents

MO4B: RFSOI Technology and Applications

Chair: Paul Hurwitz, Tower Jazz — Co-Chair: Mehmet Kaynak, IHP GmbH
Venue: Grand Salon B, 15:40 – 17:20, Monday 26 January 2015

PAGE 27
MO4B-1
RFSOI Programmable Array of Capacitors *(Invited Paper)*
(M. Granger-Jones, J. Bendixen, J. Costa, M. Carroll, D. Kerr, C. Iversen, P. Mason, E. Spears)

PAGE 30
MO4B-2
Improvements in SOI Technology for RF Switches *(Invited Paper)*
(Mark Jaffe, Michel Abou-Khalil, Alan Botula, John Ellis-Monaghan, Jeffrey Gambino, Jeff Gross, Zhong-Xiang He, Alvin Joseph, Richard Phelps, Steven Shank, James Slinkman, Randy Wolf)

PAGE 33
MO4B-3
High Resistivity SOI Wafer for Mainstream RF System-on-Chip *(Invited Paper)*
(Jean-Pierre Raskin, Eric Desbonnets)

PAGE 37
MO4B-4
Comparison of Substrate Effects in Sapphire, Trap-Rich and High Resistivity Silicon Substrates for RF-SOI Applications
(Vikram Sekar, Chih-Chieh Cheng, Richard Whatley, Chang Zeng, Alper Genc, Tero Ranta, Francis Rotella)

SiRF 2015 Table of Contents

TU1C: Topics in RF Modeling and Characterization Techniques

Chair: Hasan Sharifi, HRL Laboratories — Co-Chair: Monte Miller, Freescale
Venue: Grand Salon B, 08:00 – 09:40, Tuesday 27 January 2015

Not
Available
TU1C-1
Tunable Filters and Antennas for 4G LTE Systems (Invited Paper)
(G.M. Rebeiz, C.H. Ko, Y. Cho, B. Avser, A. Alazemi, O. Gurbuz)

PAGE 40
TU1C-2
Multitone-FM Analysis of MEMS Varactor Phase Noise Contribution in VCOs
(Gerhard Kahmen, Hermann Schumacher)

PAGE 43
TU1C-3
L-2L De-Embedding Method with Double-T-Type PAD Model for Millimeter-Wave Amplifier Design
(Seitaro Kawai, Korkut Kaan Tokgoz, Kenichi Okada, Akira Matsuzawa)

PAGE 46
TU1C-4
Cross-Line Characterization for Capacitive Cross Coupling in Differential Millimeter-Wave CMOS Amplifiers
(Korkut Kaan Tokgoz, Kimsrun Lim, Yuuki Seo, Seitaro Kawai, Kenichi Okada, Akira Matsuzawa)

SiRF 2015 Table of Contents

TU3C: Power Amplifer Applications

Chair: Julio Costa, Qorvo — Co-Chair: Paul Hurwitz, Tower Jazz
Venue: Grand Salon B, 13:30 – 14:50, Tuesday 27 January 2015

PAGE 49
TU3C-1
A +18dBm Broadband CMOS Power Amplifier RFIC with Distortion Cancellation
(Ahmed M. El-Gabaly, Carlos E. Saavedra)

PAGE 52
TU3C-2
A 1.8 to 2.4GHz Stacked Power Amplifier Implemented in 0.25μm CMOS SOS Technology
(Sultan R. Helmi, Hengying Shan, Saeed Mohammadi)

PAGE 55
TU3C-3
Channelized Active Noise Elimination (CANE) with Envelope Delta Sigma Modulation
(Rui Zhu, Yonghoon Song, Yuanxun Ethan Wang)

PAGE 58
TU3C-4
A 60GHz Highly Reliable Power Amplifier with 13dBm P$_{sat}$ 15% Peak PAE in 65nm CMOS Technology
(Boris Moret, Nathalie Deltimple, Eric Kerherve, Aurélien Larie, Baudouin Martineau, Didier Belot)

TU5A: RWS-SiRF Joint Session: RF & Internet of Things

Chair: Jeremy Muldavin, MIT Lincoln Laboratory — Co-Chair: Karen Gettings, MIT Lincoln Laboratory
Venue: Grand Salon A, 16:00 – 17:20, Tuesday 27 January 2015

Not
Available
TU5A-1
Redefining the Leading Edge: A Silicon RF Perspective *(Invited Paper)*
(P. Colestock)

PAGE 61
TU5A-2
RF and Microwave Technology Challenges for Internet-of-Things Applications *(Invited Paper)*
(L. Larson)

SiRF 2015 Table of Contents

WE1C: Tunable and Reconfigurable Technologies

Chair: J.P. Raskin, Université catholique de Louvain (UCL) — Co-Chair: Monte Miller, Freescale
Venue: Grand Salon B, 08:00 – 09:40, Wednesday 28 January 2015

PAGE 63
WE1C-1
Reconfigurable Solutions for Mobile Device RF Front-ends *(Invited Paper)*
(Arthur S. Morris III)

PAGE 67
WE1C-2
An Integrated Reconfigurable Tuner in 45nm CMOS SOI Technology
(Alice Yi-Szu Jou, Chen Liu, Saeed Mohammadi)

PAGE 70
WE1C-3
Ferroelectric MIM Capacitors for Compact High Tunable Filters
(Rosa De Paolis, Sandrine Payan, Mario Maglione, Guillaume Guegan, Fabio Coccetti)

PAGE 73
WE1C-4
10.6THz Figure-of-Merit Phase-Change RF Switches with Embedded Micro-Heater
(Jeong-Sun Moon, Hwa-Chang Seo, Dustin Le, Helen Fung, Adele Schmitz, Thomas Oh, Samuel Kim, Kyung-Ah Son, Baohua Yang)

SiRF 2015 Table of Contents

WE2C : SiRF Circuits and Applications - 2

Chair: Chiennan Kuo, National Chiao Tung University — Co-Chair: Austin Chen, Skyworks Solutions
Venue: Grand Salon B, 10:10 – 11:30, Wednesday 28 January 2015

PAGE 76
WE2C-1
Low Power and High Speed OOK Modulator for Wireless Inter-Chip Communications
(Hae Jin Lee, Chong Hyun Yoon, Joong Geun Lee, Chae Jun Lee, Dongmin Kang,
In Sang Song, Seong Jun Cho, Hong Yi Kim, Inn-Yeal Oh, Chul Soon Park)

PAGE 80
WE2C-2
A 20GHz Class-C VCO Using Noise Sensitivity Mitigation Technique
(Kento Kimura, Kenichi Okada, Akira Matsuzawa)

PAGE 83
WE2C-3
Radio-Frequency Flexible Transistors on Cellulose Nanofibrillated Fiber (CNF) Substrates
(Jung-Hun Seo, Tzu-Hsuan Chang, Ronald Sabo, Zhiyong Cai, Shaoqin Gong, Zhenqiang Ma)

PAGE 86
WE2C-4
Phase Noise Reduction in RF Oscillators Utilizing Self-Injection Locked and Phase Locked Loop
(Li Zhang, Ajay K. Poddar, Ulrich L. Rohde, Afshin S. Daryoush)

WE3P : Joint RWW Interactive Poster Session: Transceivers and Front-end Technologies SOC and SiP

Chair: Yupeng Jia, National Instruments
Venue: Grand Salon CDE, 12:50 – 14:40, Wednesday 28 January 2015

PAGE 89
WE3P-10
Gold Nanorod Array Structured Silicon Nitride Films for Reliable RF MEMS Capacitive Switches
(L. Michalas, S. Xavier, M. Koutsoureli, O. El Jouaidis, S. Bansropun, G. Papaioannou, A. Ziaei)

PAGE 92
WE3P-12
Miniaturized 60GHz Triangular CMOS Antenna-on-Chip Using Asymmetric Artificial Magnetic Conductor
(Adel Barakat, Ahmed Allam, Hala Elsadek, Adel B. Abdel-Rahman, S. Muhammad Hanif, Ramesh K. Pokharel)

2015 IEEE Radio & Wireless Week

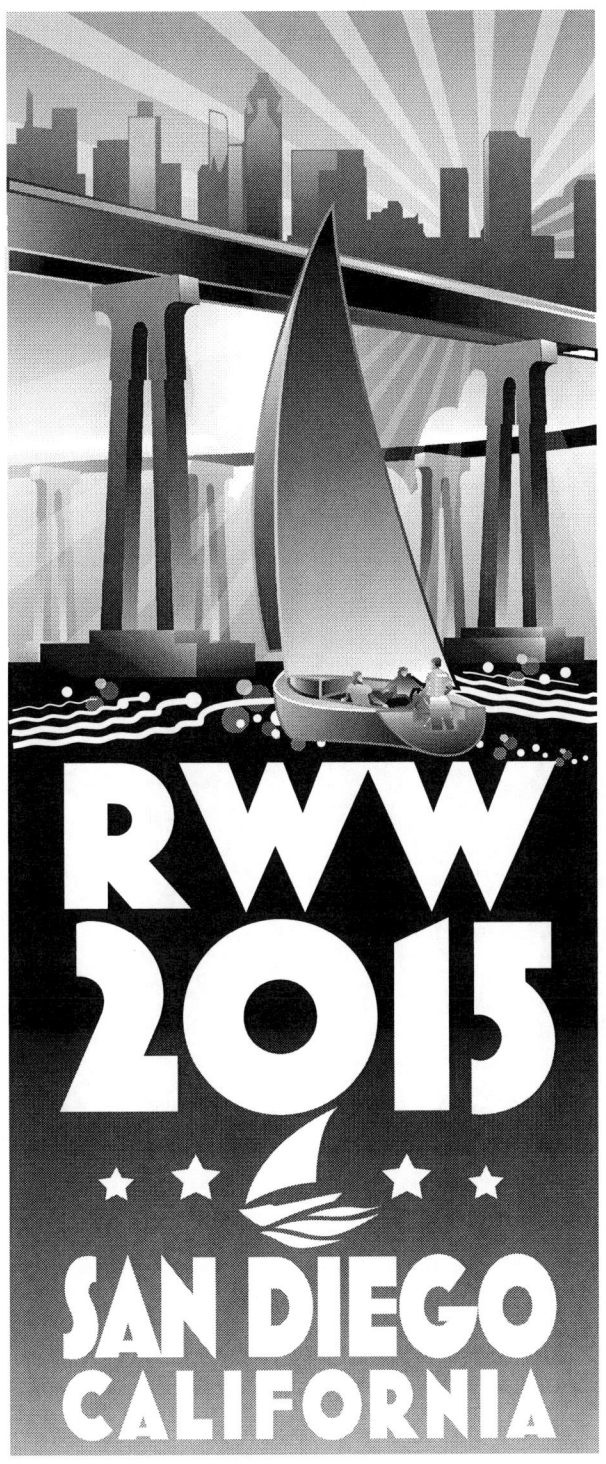

FINAL PROGRAM

Omni Hotel
San Diego, California, USA
25–28 January, 2015

RWW & RWS
General Chair:
Karl Varian

General Co-Chair:
Sergio Pacheco,
Freescale

RWW & RWS
Technical Program
Co-Chairs:
Jeremy Muldavin,
MIT Lincoln Laboratory

Mehdi Shadaram,
*University of Texas at
San Antonio*

RWW & RWS
Finance Chair:
Rashaunda Henderson,
*University of Texas at
Dallas*

WiSNet
Conference Co-Chairs:
Alexander Koelpin,
*University of
Erlangen-Nuremberg*

Rahul Khanna, *Intel*

PAWR
Conference Co-Chairs:
Almudena Suarez
Rodriguez, *University of
Cantabria*

Fred Schindler, *Qorvo*

BioWireleSS
Conference Co-Chairs:
Katia Grenier,
LAAS-CNRS

Syed Kamrul Islam,
University of Tennessee

SiRF
Conference Chair:
Chien-Nan Kuo,
*National Chiao Tung
University*

SiRF Technical
Program Chair:
Julio Costa,
Qorvo

SiRF Technical
Program Co-Chair:
Hasan Sharifi,
HRL Laboratories

RWS, PAWR, WiSNet,
BioWireleSS
Publications Chairs:
Wasif Tanveer Khan,
Spyridon Pavlidis,
Aida L. Vera Lopez
*Georgia Institute of
Technology*

SiRF
Publication Chair:
Ming-Ta Yang,
Qualcomm

2015 Radio & Wireless Week Sponsors:
IEEE Microwave Theory and Techniques Society (MTT-S)
IEEE Antennas and Propagation Society (APS)
IEEE Engineering in Medicine & Biology Society (EMBS)

http://www.radiowirelessweek.org

General Chair's Invitation to the IEEE Radio and Wireless Week

I have the great honor and pleasure to invite you to the 2015 IEEE Radio Wireless Week (RWW). This will be the ninth RWW and our third time in San Diego.

RWW2015 will be held at the Omni San Diego Hotel, San Diego, California, 25 – 28 January, 2015. The venue is nestled in the heart of the historic Gaslamp Quarter, and is just moments away from the city's top sites and attractions. With many wireless semiconductor companies, aerospace and defense industry, as well as world class universities in the area, San Diego will be a great location for all the attendees. RWW2015 consists of five related conferences that focus on the intersection between wireless communication theory, systems, circuits, and device technologies creating a unique forum for engineers to discuss various technologies for state-of-art wireless systems and their end-use applications. The conference targets to bridge the gap between digital, RF, hardware, and software that need to be seamlessly combined to keep wireless industry and mobile applications growing.

RWW's multidisciplinary events bring together innovations that are happening across the broad wireless spectrum. It is our hope that RWW is a place where you will not only find discussions or present problems, but you will also be inspired by the diverse technical content that might spark ideas for future research. The diversity of RWW is underlined by three diverse co-sponsor IEEE societies: Microwave Theory and Techniques Society (MTT-S), Antennas and Propagation Society (APS), and Engineering in Medicine and Biology Society (EMBS).

In addition to traditional podium presentations and poster sessions, there will be a track for IEEE Distinguished Lectures, Sunday and Monday half day workshops, panels, industry exhibits, WirelessApps industry presentations, and a demo session. A highlight on Tuesday will be the plenary talk on "Wearable Wireless Sensor Technologies for Truly Personalized Medicine and Wellness" by Dr. Chris Van Hoof, who is Director of Wearable Healthcare at imec in Leuven, Belgium and Eindhoven, the Netherlands. Also on Tuesday afternoon, in its third year, there will be a demo session where presenters can bring in a demonstration of their latest wireless experiments for a hands-on interactive forum. Demo sessions are particularly appropriate with the spirit of RWW because we get to see and feel how people are tackling real-world problems to address the next wireless innovation.

To support and encourage students pursuing a career in wireless area, each conference will have a student paper competition with awards that will be presented at the Tuesday banquet. On Monday afternoon, all student paper competition finalists will present their work in the poster session. I encourage you to check out what the next generation of wireless engineers are working on.

I would like to invite everyone to join us for 3 ½ days of great technical presentations, discussions, networking, and some fun in beautiful San Diego, California, 25-28 January 2015.

RWW2015 General Chair
Karl Varian

General Chair
Karl Varian

Technical Program Chair
Jeremy Muldavin

RWS 2015 Technical Program Committee

Passive Antennas
Chair: Jiang Zhu
James Schaffner   Marco Antoniades
Alessandro Cidronali   Goutam Chattopadhyay
Ahmed Kishk   Songnan Yang
Glauco Fontgalland   Arnaud Amadjikpe

Propagation Channel Modeling and Utilization
Chair: Daniel Benevides da Costa
Emery Chen   Michael Ong Lin Chuen
Changzhi Li

Transceivers and Front-end Technologies SOC and SiP
Chair: Shoichi Narahashi
Nathalie Deltimple   T.S. Jason Horng
Wasif Tanveer Khan   Telesphor Kamgaing
Hiroshi Okazaki   Xinwei Wang
Max Scardelletti

MIMO, Signal Processing and Smart Antennas
Chair: Ramya Bhagavatula
Yazhou Wang   Michael Chia
Eiji Okamoto   Dimitris Toumpakaris
Chau Yuen

High-speed and Broadband Wireless Technologies
Chair: Shilong Pan
Beatrice Cabon   Yik-Chung Wu
Idelfonso Tafur Monroy   Minoru Fujishima

Software Defined Radios and Cognitive Radios
Chair: Abbas Omar
Nuno Borges Carvalho   Yves Baeyens
Dimitrie C. Popescu   Lin Song
Ed Niehenke

Wireless Systems Architecture and Modeling
Chair: Markos Anastasopoulos
Ugo Dias   Hyun Kyu Chung
Vegas Olmos

Emerging Wireless Technologies and Applications
Chair: Sergio Pacheco
Debabani Choudhury   Chia-Chan Chang
Dimitrios Peroulis   Yoshihiro Kawahara
Zhen Ning Low

Digital Signal Processing as Applied to Wireless
Chair: Karl Molnar
Upkar Dhaliwal   Shin Hara
Swami Sankaran   Renato Negra
Xinwei Wang

Passive Components and Packaging
Chair: Rashaunda Henderson
Hualiang Zhang   Roberto Gomez-Garcia
Xun Gong   Dariush Mirshekar
Clemens Ruppel

Late News Papers
Chair: Sergio Pacheco
Takao Inoue   Karl Varian
Xun Gong   Charlie Jackson
Kevin Chuang   Telesphor Kamgaing

Invited Papers
Chair: Telesphor Kamgaing
Sergio Pacheco

RWW 2015 Steering Committee

General Chair:
Karl Varian, Raytheon, retired
General Co-Chair:
Sergio Pacheco, Freescale
Technical Program Chair:
Jeremy Muldavin, MIT Lincoln Laboratory
Technical Program Co-Chair:
Mehdi Shadaram, University of Texas at San Antonio
Topical Conference PAWR Co-Chairs:
Almudena Suarez Rodriguez, University of Cantabria
Fred Schindler, Qorvo
Topical Conference BioWireleSS Co-Chairs:
Katia Grenier, LAAS-CNRS
Syed Kamrul Islam, University of Tennessee
Topical Conference WiSNet Co-Chairs:
Rahul Khanna, Intel
Alexander Koelpin, University of Erlangen-Nuremberg
SiRF General Chair:
Chien-Nan Kuo, National Chiao Tung University
Finance Chair:
Rashaunda Henderson, University of Texas at Dallas
Web Master:
Min Hua, Raysilica
Workshops Chair:
Nuno Borges Carvalho, Universidade de Aveiro
Plenary & Panel Session Chair:
Rizwan Bashirullah, University of Florida
Distinguished Lectures Session Chair:
Hermann Schumacher, Ulm University
Poster Session Chair:
Yupeng Jia, National Instruments
Demo Track Chair:
Changzhi Li, Texas Tech University
Paper Submission Management System Chair:
Kevin Chuang, MIT Lincoln Laboratory
Publications Co-Chairs:
Wasif Tanveer Khan, Georgia Insitute of Technology
Spyridon Pavlidis, Georgia Institute of Technology
Aida L. Vera Lopez, Georgia Institute of Technology
MicroApps Chair:
Sherry Hess, AWR
Student Paper Awards Co-Chairs:
Abbas Omar, University of Magdeburg
Holger Maune, Technical University of Darmstadt
Publicity Co-Chairs:
Li Lu, Qualcomm
Local Publicity and Coordination Chair:
Madhu Gupta, San Diego State University
Microwave Magazine Special Issue Editor:
Dietmar Kissinger, IHP GmbH
Exhibition/Sponsorship Co-Chairs:
Charlie Jackson, Northrup Grumman Corp.
Mark Hoffman, AceTech, Inc.
Conference Management:
Elsie Cabrera, IEEE
International Liaison:
Zaher Bardai, IEEE
RWW Executive Committee Chair:
Charlie Jackson, Northrup Grumman Corp.
Special Consultant and Adviser:
John Barr, IEEE MTT Society
George Heiter, Heiter Microwave Consulting
Takao Inoue, National Instruments
Donald Lie, Texas Tech University
Jan-Erik Mueller, Intel

The 15TH Topical Meeting on Silicon Monolithic Integrated Circuits in RF Systems

Message from the SiRF General Chair:

Welcome to SiRF 2015!

The IEEE Topical Meeting on Silicon Monolithic Integrated Circuits in RF Systems (SiRF) continues moving forward to its 15th year. The establishment was inspired by the fast-growing capability of full system integration in silicon (Si) technology, combining high-speed digital and RF circuitry as a total solution. The early years of this conference were a time of pursuing for device and circuit performance in Si-based design. The advance of Si technology kept successfully demonstrating impressive progress, and steadily spreading the application spectrum into the millimeter-wave regime and higher.

It is now generally acknowledged that CMOS monolithic microwave integrated circuits (MMICs) is catching up with GaAs semiconductor technologies to deliver sufficient efficiency and fulfill the general requirement in a variety of commercial RF products that used to be a forbidden area to Si technology. The technical scope of this conference therefore has been greatly extended to a broad range of technology development and system applications. Device technologies include smart materials, nano-technologies, narrowwire and grapheme, Si photonics, and so on. Circuit and system applications cover designs of mixed-signal, microwave, millimeter-wave, and terahertz (THz) frequency bands, based on Si-related technology. System characterization takes into account heterogeneous System-in-Package (SiP), flexible electronics, and sensor systems. The critical development of SiGe BiCMOS and RF silicon-on-insulator (RFSOI) has been embedded in conversations in the conference venue. The conference offers an invaluable platform networking worldwide RF IC designers and researchers for experience sharing of amazing breakthroughs and dialogues of the future trends. SiRF2015 will continue as a part of Radio and Wireless Week (RWW) on Jan. 26-28 2015 in the harbor city of San Diego in Southern California, United States, where it participated RWW the first time. SiRF2015 is sponsored by IEEE Microwave Theory and Techniques Society (MTT-S).

To address the emerging technology and future direction of research and development, several reputed speakers are invited from the academia and industry to nourish conversations. The partial list of tentative speakers is as follows.

1. Prof. Erich Kasper, Stuttgart University, will present silicon monolithic millimeter-wave integrated circuit integration of millimeter-wave antenna with two terminal devices for medical applications.
2. Prof. Huei Wang, National Taiwan University, will discuss millimeter-wave IC design in CMOS technology.
3. Prof. Gabriel Rebeiz, University of California San Diego, will share research of tunable filter design.
4. Prof. Larry Larson, Brown University, will talk about the applications and challenges of Internet of Things.
5. Dr. Paul Colestock, Global Foundries, will show the roadmap of technology development.
6. Dr. Art Morris, Wispry, will present the company core technology of tunable/reconfigurable MEMS capacitors integrated into RFCMOS.

Furthermore, it is worth mentioning that in this year's conference a great focus session will be organized on topics related to RFSOI technology. Several talks from Qorvo, IBM, Tower Jazz and other companies will reveal current updates and future trends to the interested audience.

You are welcome to join us at SiRF2015 in January 2015. Authors please demonstrate your work summarized in a three-page manuscript in PDF format. Note that selected conference papers will be considered for publication in IEEE Transactions on Microwave Theory and Techniques with a significant extension through the regular review process.

Please visit us at http://www. silicon-rf.org to see further details. We will be meeting each other in San Diego!

Yours sincerely,

Chien-Nan Kuo, Ph. D.
National Chiao Tung University
cnkuo@mail.nctu.edu.tw
SiRF 2015 General Chair

SiRF 2015 Technical Program Committee

Technical Program Chairs:
Julio Costa, Qorvo
Hasan Sharifi, HRL Laboratories

Technology, Devices and Modeling
Chair: Mehmet Kaynak, IHP GmbH
Julio Costa, Qorvo
Mingta Yang, Qualcomm
Katsuyoshi Washio, Hitachi
Guofu Niu, Auburn University
Harrie Tilmans, IMEC
Paul Hurwitz, Tower Jazz

Passives and MEMS
Chair: Jean-Pierre Raskin, UCL
Xun Gong, UCF
Pierre Blondy, University Limoges
Emmanuel Defay, LETI
Hasan Sharifi, HRL Laboratories

Circuits
Chair: Larry Larson, Brown University
Hermann Schumacher, Ulm University
Vince Fusco, Queens University of Belfast
Lance Kuo, Raytheon
Yunliang Zhu, Qualcomm
Austin Ying-Kuang Chen, Skyworks Solutions
Hsieh-Hung Hsieh, TSMC
Kenichi Okada, Tokyo Inst. of technology
Monte Miller, Freescale
Gang Liu, University of California San Diego

Applications and Wireless Architectures
Chair: Francesco Dantoni, TI
Donald Y. C. Lie, Texas Tech University
Chien-Nan Kuo, NCTU
Jürgen Hasch, Bosch
Yan Li, Qorvo
Himanshu Khatri, Qualcomm

Late News Papers
Chair: Sergio Pacheco, Freescale
Takao Inoue, National Instruments
Karl Varian, Raytheon
Xun Gong, University of Central Florida
Charlie Jackson, Northrop Grumman
Kevin Chuang, MIT Lincoln Laboratory
Telesphor Kamgaing, Intel

REGISTRATION HOURS

Registration is open during the following times at the Grand Ballroom Foyer:

Sunday, 25 January: 12:00-17:00
Monday, 26 January: 07:00-19:00
Tuesday, 27 January: 07:00-17:00
Wednesday, 28 January: 07:00-12:00

EXHIBIT HOURS

The exhibition area (Salons C,D,E) is open during the following times:

Monday, 26 January 2015 13:00 – 17:30
Tuesday, 27 January 2015 10:00 – 17:00

For the latest information and details on how to become a sponsor and exhibit at RWW please visit: http://www.radiowirelessweek.org/exhibits.

SiRF 2015 Steering Committee

General Chair:
Chien-Nan Kuo, National Chiao Tung University
Technical Program Co-Chairs:
Julio Costa, Qorvo
Hasan Sharifi, HRL Laboratories
Publicity Chair:
Xun Gong, University of Central Florida
Publications Chair:
Ming-Ta Yang, Qualcomm
International Liaison Europe:
Dietmar Kissinger, IHP GmbH
International Liaison Asia:
Jae-Sung Rieh, Korea University
Executive Committee:
Mehmet Kaynak, IHP GmbH
Dietmar Kissinger, University of Erlangen-Nuremberg
Chien-Nan Kuo, National Chiao Tung University
Rudolf Lachner, Infineon Technologies AG
Donald Y. C. Lie, Texas Tech University
Zhenqiang (Jack) Ma, University of Wisconsin-Madison
Guofu Niu, Auburn University
Sergio Pacheco, Freescale Semiconductor
Dimitrios Peroulis, Purdue University
Nils Pohl, Ruhr-Universität Bochum
George Ponchak, NASA Glenn Research Center
Jae-Sung Rieh, Korea University
Clemens Ruppel, EPCOS AG
Hermann Schumacher, Ulm University
Vaclav Valenta, Ulm University
Katsuyoshi Washio, Tohoku University
Robert Weigel, University of Erlangen-Nuremberg

SOCIAL EVENTS

Complimentary Daily Breakfast (Mon.-Wed.)
Place: Palm Terrace
Time: 07:00-08:00

Complimentary Daily AM Coffee Breaks
Place: Salon CDE
Time: 9:40-10:10

Complimentary Daily PM Coffee Breaks
Place: Salon CDE
Time: 15:10-15:40

RWW Reception
Place: Palm Terrace
Monday 18:00-20:00

RWW/SiRF Awards Banquet
Place: Gallery 3B
Tuesday 18:30-21:00

RWW Topical Conferences

Power Amplifiers for Radio and Wireless Applications (PAWR)

Interest in power amplifier technology remains at an all time high because of the emergence of new device materials such as GaN that offer improved performance, and the need for ever greater linearity and efficiency by the world's expanding wireless communication infrastructure. This year, the Topical Conference on Power Amplifiers for Wireless and Radio Applications (PAWR) will feature a full day of power amplifier focused sessions, including the latest advances on power amplifier technology, efficiency enhancement techniques, system analysis, modeling and distortion reduction, an interactive workshop on Advances on Power Amplifiers for Modern Wireless Communications and a panel session on 5G Power Amplifier Technologies.

Technical Committee:

Distortion Reduction Techniques in RF Power Amplifiers
Chair: Allen Katz
Slim Boumaiza   Armando Cova
Kiki Ikossi   Jinsung Choi
Peter Kenington   Shabbir Moochalla
Timo Rahkonen   Joe Staudinger

High Efficiency RF Power Amplifiers
Chair: Arturo Mediano
James Komiak   Song Lin
Chao Lu   Mohammad Madihian
Frederick Raab   Dave Runton
Ali Tombak   John Walker

RF Power Amplifier Technology
Chair: Marc Franco
Nick Cheng   Nathalie Deltimple
Murat Eron   Gary Hau
Bumman Kim   Donald Lie
Zoya Popovic   Paolo Colantonio

Power Amplifier Modeling and System Analysis
Chair: Andrei Grebennikov
Francis Rotella   Robert Caverly
Almudena Suarez   Wolfgang Heinrich
Stephen Maas   Jose Carlos Pedro
Gene Tkachenko

Wireless Sensors and Sensor Networks (WiSNet)

WiSNet is dedicated to the advancement of wireless sensors for commercial and industrial applications and will be held to specifically focus on the latest developments in these areas of RF Sensors and Sensor Networks. Wireless sensors and sensor networks are critical system components for manufacturing, monitoring, safety, as well as positioning and tracking applications. This year, WiSNet2015 will be a full day topical conference focused on the latest developments in these areas. Different sessions will focus on sensors and smart sensor networks ranging from UHF, RFID applications to millimeter-wave radar systems.

Technical Committee:

Wireless Sensors for Communication, Radar, Positioning and Imaging Applications
Chair: Martin Vossiek
Changzhi Li   Aly Fathy
Mario Pauli   Kamal Samanta

Wireless Sensors for Localization, Tracking, and RFID Technologies
Chair: Manos M. Tentzeris
Xianming Qing   Apostolos Georgiadis
Hao Xin   Reinhard Feger

Wireless Integrated Sensors, Front-Ends, and Building Blocks
Chair: Linus Maurer
Huei Wang   Thomas Ussmueller
Nils Pohl   Andreas Baenisch
Daniela Dragomirescu   Holger Maune

Wireless Sensors for Harsh Environments, Home, Health and Communication
Chair: Alexander Koelpin
Georg Fischer   Arne Jacob
Maurizio Bozzi   Hendrik Rogier

Sensor Network Communication Architecture and Topologies
Chair: Rahul Khanna
Alexander Koelpin   Xun Gong
Huaping Liu

Multi-port Technology
Chair: Alexander Koelpin
Serioja Tatu   Iñigo Molina Fernández
Fadhel Ghannouchi   Tuami Lasri
Adriana Serban   Gabor Vinci

Wireless Sensors for Wearable Computing and Internet of Things
Chair: Rahul Khanna
Alexander Koelpin   Xun Gong
Huaping Liu

Invited Papers
Chair: Rahul Khanna
Alexander Koelpin

Biomedical Wireless Technologies, Networks, and Sensing Systems (BioWireleSS)

The IEEE Topical Conference on Biomedical Wireless Technologies, Networks, and Sensing Systems (BioWireleSS) will be a vital part of the IEEE Radio and Wireless Symposium, featuring the latest developments in wireless biomedical technologies, networks and sensing systems. The wireless revolution has begun to infiltrate the medical community with patient health monitoring, telesurgery, mobile wireless biosensor systems, and wireless tracking of patients and assets becoming a reality. The rapid evolution of wireless technologies coupled with powerful advances in adjacent fields such as biosensor design, low power battery operated systems, and diagnosing and reporting for intelligent information management has opened up a plethora of new applications for wireless systems in medicine.

Technical Committee:

Wireless Technologies for Biosignals and Modeling in Medical Environments
Chair: Jung-Chih Chiao
Alper Bozkurt   Natalia Nikolova
Marc Notten   Mohammad-Reza Tofighi
Aydin Farajidavar   Nicole McFarlane
Jeremy Holleman

Wireless Position and Localization in Medicine
Chair: Changzi Li
Andreas Stelzer   David Ricketts
Michael Kuhn   Upkar Dhaliwal
Aydin Farajidavar   Mohamed Mahfouz
Aly Fathy

PAN, BAN, Energy Scavenging and Remote Patient Monitoring
Chair: Changzhi Li
Dietmar Kissinger   Dominique Schreurs
David Ricketts   Yong Xin Guo
Syed Islam   Aydin Farajidavar

Micro-Sensors and In-vivo Microsystems
Chair: David Dubuc
Jung-Chih Chiao   Marc Notten
Michael Kuhn   Arnaud Pothier
Claire Dalmay   Katia Grenier
Alper Bozkurt   Syed Islam
Melika Roknsharifi   Joachim Oberhammer
Pingshan Wang   Hung-Wei Wu
Rizwan Bashirullah

Microwaves in Biological Applications and Interaction with Biological Tissues
Chair: Mohammad-Reza Tofighi
Yong Xin Guo   Victor Lubecke
Dominique Schreurs   Indira Chatterjee
Usmah Kawoos   Andre Vander Vorst
Katia Grenier   Jung-Chih Chiao
Arye Rosen   David Dubuc
Joachim Oberhammer

Medical Imaging and Applications
Chair: Natalia Nikolova
Arye Rosen   Usmah Kawoos
Anand Gopinath   Victor Lubecke
Changzhi Li   Mohammad-Reza Tofighi
Bashir Morshed

Invited Papers
Chair: Katia Grenier
Syed Islam

Focused Sessions & Others
Chair: Syed Islam
Katia Grenier

Diamond Sponsor:

Technical Program for 2015 Radio Wireless Week (RWW)

WORKSHOPS/INDUSTRY FORUM/PANEL

SUNDAY, 25 JANUARY 2015 (13:30-17:30)

Workshop **Millimeter Waves in 5G: State of the Art and Potential**	*Workshop* **RFID Technologies**	*Workshop* **3D Printing and its Impact on Wireless Systems**	*Workshop* **Advances on Power Amplifiers for Modern Wireless Communications**
Room: Gaslamp 1	Room: Gaslamp 2	Room: Gaslamp 3	Room: Gallery 1

Organizer:
Wilhelm Keusgen, Fraunhofer Heinrich-Hertz-Institute, Berlin, Germany

The millimeter-wave frequency band is seen as a good candidate for future 5G mobile radio networks. The availability of multiple Gigahertz of bandwidth promises to be an answer to the ever increasing data rates in access and backhaul links. At the same time new challenges arise. Mass production of RF communication circuits for this band is at its beginning. The knowledge on the outdoor channel is still very limited. New approaches for integrated steerable high gain antennas are necessary to counter the higher path loss at the high frequencies. The network architecture is expected to become more and more heterogeneous with a mixture of different cell radii and the wide deployment of small cells.

With views from industrial and scientific research this workshop will give an overview on the state of the art and the huge potentials of millimeter-waves in 5G.

Organizers:
Thomas Ussmueller, University of Innsbruck, Austria
Apostolos Georgiadis, CTTC, Spain

Radio-frequency identification (RFID) is a technology for wireless communication and sensing. Most of today's RFID tags are passive tags without their own power supply. Thus they have to rely on the electro-magnetic energy of the incoming data signal or on other sources of energy for powering the the tag.

This workshop is a tutorial workshop discussing the basic principles of RFID systems. The first talk will give an overview on RFID technologies. It will discuss the various available frequency bands their physical properties and their typical usage scenarios. In addition the talk will cover the different available RFID standards. The second talk will focus on wireless powering and the performance of the rectifier stages responsible for powering the RFID tag chip and consequently determining to a large extent the operating range of passive RFID tags. Subsequent talks will describe the communication principles of RFID systems and provide a summary of existing as well as potential applications of the technology.

Organizer:
Manos M. Tentzeris, Georgia Institute of Technology, USA

This workshop will be focused on alternative solutions to implement 3D circuit design and inkjet printing for new and emerging wireless systems. The speakers will focus their talks on inkjet printed wireless sensor networks and smart 3D surfaces.

Organizer:
Slim Boumaiza, University of Waterloo, Canada

This workshop will be devoted to the design of Power Amplifiers for Modern Wireless Communications, spanning from Next generation Doherty to outphasing and special techniques for wideband amplifier.

Speakers:

Millimeter-Waves for Future Mobile Communications
Wilhelm Keusgen, Fraunhofer Heinrich-Hertz-Institute, Germany

Hybrid Precoding and Channel Estimation Algorithms for Millimeter-Wave Systems
Robert W. Heath Jr., The University of Texas at Austin, USA

Performance evaluation of 5G cellular networks with millimeter-wave small-cell base stations
Kei Sakaguchi, Tokyo Institute of Technology and Osaka University, Japan

Channel Measurement and Modeling for Millimeter-Wave Mobile Communication
Richard Weiler, Fraunhofer Heinrich-Hertz-Institute, Germany

Hardware Realizations to Millimeter-Wave 5G Systems
Tauno Vähä-Heikkilä,
VTT Technical Research Centre of Finland, Finland

mmWave Technology Evolution for Next Generation Wireless Systems
Ali Sadri, Intel Corporation, USA

Unleashing Millimeter-Wave Frequencies – Test and Measurement Aspects
Andreas Rößler , Rohde & Schwarz USA Inc., USA

Speakers:

Overview of RFID standards
Thomas Ussmueller, University of Innsbruck, Austria

Wireless Power Transmission for RFID-tags
Apostolos Georgiadis, Centre Tecnologic de Telecomunicacions de Catalunya (CTTT), Spain

Wireless Data Transmission in RFID Systems
Thomas Ussmueller, University of Innsbruck, Austria

RFID Applications: present, future and futuristic ones
Luca Roselli[1] and Alessandra Costanzo[2], [1]University of Perugia, [2]University of Bologna, Italy

SAW RFID-Transponder-based wireless systems and applications
Amelie Hagelauer, University of Erlangen-Nuremberg, Germany

Speakers:

Additive Manufacturing Techniques for RF Modules and WSN's
Manos Tentzeris, Georgia Tech, USA

Inkjet pritned antennas and circuits for energy harvesting and wireless sensors
Apostolos Georgiadis, CTTC, Spain

A novel reconfigurable origami accordion antenna
Benjamin Cook, Georgia Tech, USA

System in Package on Paper (SiPoP) technology as a means to realize extremely low cost 3D millimeter-wave circuits and systems
Luca Roselli, Perugia University, Italy

Smart Surfaces using WPT
Nuno Borges Carvalho, DETI – Instituto de Telecomunicações, Universidade de Aveiro, Portugal

Speakers:

Title: Next Generation Doherty and Outphasing Amplifiers
Leo de Vreede, University of Delft, Netherlands

The Doherty Power Amplifier for Broadband or Multiband Communication Systems
Paolo Colantonio, University of Roma Tor Vergata, Italy

Broadband Doherty Amplifier and Envelop Tracking Power Amplifier for Carrier Aggregated Signals
Slim Boumaiza, University of Waterloo, Canada

Varactor Based Dynamic Load Modulation Amplifiers for Wideband and Multiband Applications
Christian Fager, Chalmers University, Sweden

REGISTRATION

Advance registration for RWW 2015 is open now until January 5, 2015. Register now to take advantage of the early registration pricing! Please note that workshop fees are additional.

Please visit http://www.radiowirelessweek.org/attendees/registration-information/ for more information.

MONDAY, 26 JANUARY 2015

Workshop
Microwave Biosensing Developments in Asia

Time: 13:30-17:30
Room: Gaslamp 1

Organizer:
Hung-Wei Wu, Kun Shan University, J.-C. Chiao, University of Texas - Arlington, USA

In recent years, the development of advanced RF/microwave/wireless sensing techniques for emerging biomedical applications have made significant progresses and shown great promises for commercial uses and improving human well-being. Innovative technologies and integration of electromagnetics and biology have opened new opportunities for scientific researches and healthcare applications globally. The aims of this workshop are to report recent research achievements by invited experts from Asia and motivate interactive discussions among international attendees in the promising areas of RF, microwave and wireless biosensing. The workshop focuses on aspects of wireless sensor devices, bio-signals, bio-materials, biochemical sensing, wireless power transfer and vital sign detection.

Attendees in this workshop will be able to:
1. Obtain a broad, state-of-the-art overview of materials, devices, systems and measurement techniques for microwave biosensing technologies in five different regions;
2. Comprehend in-depth knowledge and technical obstacles in industry standards compliance, innovative research, and practical scenarios with sharing of firsthand experience from these experts;
3. Discuss critical issues and technical challenges facing in laboratories and commercialization;
4. Inspire new research ideas and share synergistic concepts among the global RF, microwave and wireless communities.

Speakers:

Wireless Sensor Microsystems for Medical Devices
Prof. Minkyu Je, Daegu Gyeongbuk Institute of Science & Technology (DGIST), Korea

Wireless Sensing and Measurement of Doppler Bio-signals and Bio-materials
Prof. Lixin Ran, Zhejiang University, China

State-Of-The-Art of Wireless Technologies for Medical Bio and Biochemical Sensing in Asia and Critical Aspects of Such Technologies
Prof. Agnes Tixier-Mita, University of Tokyo, Japan

Wireless Power Transfer Technology for Medically Implantable Devices
Prof. Franklin Young-Jae Bien, Ulsan National Institute of Science and Technology (UNIST), Korea

A Review on Microwave/Millimeter-wave Sensor Systems for Vital Sign Detection
Prof. Huei Wang, National Taiwan University, Taiwan

Industry Forum
Tutorial Workshop on Advances in SiGe BiCMOS Technology with Chip Scale Phased Array Applications

Time: 09:00 - 12:00
Room: Gallery 3A

Organizer and Speaker: Gabriel M. Rebeiz, Wireless Communications Industry Endowed Chair Professor, University of California-San Diego, USA

This course will present the latest work on microwave and mm-wave phased arrays at UCSD and selected companies. The course will show that one can build large phased arrays on a single chip covering distinct frequency bands, from 2 GHz to > 94 GHz, using commercial CMOS and SiGe processes. The course will start with some on-chip phased array architectures and the pros and cons of each architecture. Typical designs include an 8-element 8-16 GHz SiGe phased array receiver, a 16-element Tx/Rx phased array at 42-48 GHz with 5-bit amplitude and phase control, 8, 16 and 32-element 60 GHz phased array chips from industrial contributors, 16- element Rx phased array at 77-84 GHz which includes a built-in-self-test system Also, an 8-20 GHz digital beam-former chip capable of multiple beam operation and with high immunity to interferers will also be presented. In terms of wafers-scale designs, 94 GHz and 110 GHz wafer-scale phased arrays will also be presented including high efficiency antennas. The course will conclude with packaging techniques for highly dense phased arrays which are as critical as the chip itself since packaging can have a severe effect on the coupling between the channels. It will be shown that SiGe and CMOS has changed the way we think about phased arrays and has allowed the fabrication of highly complex systems at a cost reduction of 5-10x compared to an all-GaAs solution. Most importantly, it made phased arrays a possibility/reality for a large range of low/medium cost phased arrays, such as point-to-point communications, SATCOM, and low power radars.

Panel
5G Power Amplifier Technologies

Time: 19:00-20:30
Room: Grand Salon B

Organizer: Andrei Grebennikov, Microsemi Co., USA

Moving towards the forthcoming 5G cellular systems calls for wider modulation bandwidth, higher frequencies, and high density local-area networks with a merge of many technologies and techniques. This imposes the strong requirements to power amplifiers as key elements of the cellular transmitters, both in handsets and base stations, including their reconfigurability and varactor tuning capability, carrier aggregation and linearity, frequency bandwidths up to millimeter waves, efficiency and integration level for small-cell base stations, Doherty, envelope tracking, or outphasing configurations, CMOS vs. GaAs HBT and GaN HEMT vs. LDMOSFET technologies. There will be no formal presentation, with the main emphasis to provide expert answers to questions posed by attendees, who are strongly encouraged to participate in the discussion and express their vision. Any power amplifier architectures and techniques, technologies and frequencies are open for discussion.

Moderators:
Andrei Grebennikov, Microsemi Co., USA
Murat Eron, Wireless Telecom Group Inc., USA

Panelists:
Marc Franco, Qorvo, USA
Florinel Balteanu, Skyworks, USA
James Wong, Huawei, China
Peter Asbeck, University of California-San Diego, USA
Christian Fager, Chalmers University of Technology, Sweden
Rik Jos, NXP, The Netherlands

Attractions in San Diego, CA

The 2015 IEEE Radio and Wireless Week (RWW) will be held at the Omni Hotel in San Diego, California. San Diego is a premier tourist destination for individuals and families from around the world. In addition to the options of rental cars and taxis, local transportation includes trolley and bus services, surfrider trains, and tourist buses offering day tours. The following are some of the suggested activities, organized by their distance from the conference and how you might get there:

Less than 10 min by foot	Less than 15 minutes by bus or taxi	20-60 miniutes by car	Day-long activities
Gaslamp District: This downtown business and commercial area boasts numerous restaurants, bars, shopping outlets, and nightclubs. **Petco Park:** The baseball stadium is the home of the local team, the San Diego Padres. **Ship Museums of San Diego:** The USS Midway Museum is housed on an aircraft carrier, and the Mari- time Museum of San Diego is housed in the Star of India, one of the world's oldest sailing ships. Both museums are located on North Harbor Drive, along the waterfront. **Seaport Village:** Enjoy a stroll through this park on the ocean front, with shopping, restaurants, and picnic spots to enjoy along the way. **Embarcadero:** A marina park, going south along the water, past the Coronado bridge, Embarcadero has views of the ocean and downtown as well as restaurants. **Diego Civic Theater:** Home to San Diego opera and Broadway shows, is located on Third and B streets downtown.	**Balboa Park:** It is the largest urban cultural park in the nation. It is located at the northwest corner of the downtown area. It is home to the world-famous San Diego Zoo and other beautiful gardens, such as the Japanese Friendship Garden, Botanical Garden, and Lily Pond. It also includes a collection of more than a dozen major museums, including the Natural History Museum, Air and Space Museum, Museum of Photographic Art, and San Diego Museum of Art. It also houses performing arts venues, including the Old Globe Theater. **Old Town San Diego State Historic Park:** Celebrate the Mexican heritage in California at this park with its collection of exhibits, historic sites, and entertainment outlets along with many restaurants, snack shops, and specialty shops. **Mission San Diego de Alcala:** This historic site is one of the seven original Jesuit missions, founded in 1769 and built before California became a U.S. state.	**Point Loma:** Travel to this lookout point above San Diego Bay to take photos. It is the home of Cabrillo National Monument and Old Point Loma Light House. The Fort Rosecrans National Cemetery is on the way along Cabrillo Memorial Drive. **Sunset Cliffs Natural Park:** A great place to watch the sun set over the Pacific Ocean. **Mission Bay Park:** A 4,600-acre aquatic park dedicated to leisure and active sports with kayaking opportunities. **La Jolla Coves:** A famous diving, swimming, and snorkeling spot. Closeby are Shell Beach Tide Pools and the Museum of Contemporary Arts. **Torrey Pines Gliderport:** On top of a ocean-facing cliff, watch people paragliding around the cliffs or try hang gliding yourself. **Birch Aquarium at Scripps:** Perched on a bluff overlooking the Pacific ocean, the aquarium is a public exploration center at Scripps Institution of Oceanography at the University of California, San Diego, with a large aquarium of cold-water fish.	**Sea World:** Sea World is a one-of-a-kind theme park with sea-life and dolphin shows along with an aquarium, rides, and entertainment. **LEGOLAND:** This theme park in Carlsbad, about 30 miles north of downtown, has exhibits and activities aimed at the young. **Tijuana, Mexico:** Tijuana is San Diego's neighbor across the border in Mexico (but only a half hour from downtown). It is possible to drive up to the border, park, and take a bus across the border for a walk along Revolution Boulevard. **San Diego Zoo Safari Park (or Wild Animal Park):** A 1,800-acre zoo near Escondido, it is located one-half-hour north of San Diego and features a train safari, shows, activities, and dining.

Petco Park, home to the Padres MLB franchise, is located next to the Omni Hotel
Courtesy: Omni Hotel, San Diego

MONDAY, 26 JANUARY 2015

RWW Session: MO1A	SiRF Session: MO1B	PAWR Session: MO1C	RWS Session: MO1D
RWW Distinguished Lectures I	**SiRF Circuits and Applications - 1**	**Distortion Reduction Techniques in RF Power Amplifiers**	**High-speed and BroadBand Wireless Technologies**
Chair: Hermann Schumacher, Ulm University	Chair: Larry Larson, Brown University Co-Chair: Rahul Kodkani, Qual-Comm	Chair: Allen Katz, Linearizer Technology, Inc. Co-Chair: Kiki Ikossi	Chair: Mehmet Kaynak, IHP GmbH
Room: Gallery 2	Room: Grand Salon A	Room: Gallery 1	Room: Grand Salon B

08:00

MO1A-1 An Introduction to Software Defined Radio for Engineers

Jeffrey Pawlan, Pawlan Communications

Co-Sponsored by IEEE MTT-S

Abstract: Software Defined Radio (SDR) is the culmination of advances on several fronts and probably the most significant area of development in radio systems today. The entire worldwide cellular system uses SDR. NASA and the US military communications are now almost exclusively using SDR. Soon new automobile radios will be SDR to accommodate multiple modulation formats. This lecture will begin with the definition, history and evolution of (SDR). RF/microwave engineers will find it clear and understandable because analogies will be made to conventional classic radio systems and components. A live demonstration of SDR will be presented.

MO1B-1 SIMMWIC Integration of Millmeter-Wave Antenna With Two Terminal Devices For Medical Applications (Invited)

E. Kasper, W. Zhang, University of Stuttgart, Institute of Semiconductor Engineering (IHT), Stuttgart, Germany

---

MO1C-1 Linearizers - Distortion Reduction in High Power Amplifiers (Invited)

A. Katz, The College of New Jersey/Linearizer Technology, Inc., Ewing Township, United States

---

MO1D-1 Novel Non-Square, Gray Coded, 64-QAM Constellations

D. H. Morais, Adroit Wireless Strategies, San Mateo, United States

08:20

MO1D-2 BCH and LDPC Coded Wideband Modem for 21-GHz Band Satellite Broadcasting System

Y. Suzuki, Y. Matsusaki, M. Kamei, A. Hashimoto, T. Kimura, S. Tanaka, T. Ikeda, NHK, Setagaya-ku, Japan

08:40

MO1A-2 RF Aspects of Magnetic Resonance Imaging

Robert Caverly, Villanova U.

Co-Sponsored by IEEE MTT-S

Abstract: Magnetic Resonance Imaging (MRI) scanners are an important diagnostic tool for the medical practitioner. MRI provides a non-invasive means of imaging soft tissues and to obtain real-time images of the cardio-vascular system and other dynamic changes in the human body. MRI scanners rely heavily on a number of topical areas of interest to Electrical Engineers: image processing, high speed computing and RF (radio frequency) systems and components. This presentation will focus on some of the RF aspects of the MR process and MR scanners.

MO1B-2 A SiGe Differential 50ps Gaussian Pulse Generator for Sub-Sampling TDR Measurements

G. Hasenaecker[1], H. Rein[1], K. Aufinger[2], N. Pohl[3], T. Musch[1], [1]Ruhr-Universitaet Bochum, Bochum, Germany, [2]Infineon Technologies, Neubiberg, Germany, [3]Fraunhofer FHR, Wachtberg, Germany

---

MO1C-2 A Novel Input Matching Topology for Improved Digital Pre Distortion of RF Power Devices

R. J. Wilson[1,3], S. Goel[1], P. Singer[2], [1]Infineon Technologies, Morgan Hill, United States, [2]Infineon Technologies, Villach, Austria, [3]Cardiff University, Cardiff, United States

---

MO1D-3 Compact Mono-Static/Bi-Static UWB System for Wall Parameters Extraction

S. Magoon[1], C. Thajudeen[1], A. Hoorfar[2], A. E. Fathy[1], [1]The University of Tennessee, Knoxville, Knoxville, United States, [2]Villanova University, Villanova, United States

09:00

MO1B-3 A 6.5 mW, Wide Band Dual-Path LC VCO Design With Mode Switching Technique in 130 nm CMOS

J. Li[1,2], N. Xu[1], Y. Sun[1,3], W. Rhee[1], Z. Wang[1], [1]Tsinghua University, Beijing, China, [2]University of California, San Diego, San Diego, United States, [3]Hua-Chuang Securities Brokerage CO LTD, Beijing, China

---

MO1C-3 A digital predistortion method based on nonuniform memory polynomial model using interpolated LUT

X. Feng[1], B. Feuvrie[1,2], A. S. Descamps[1,2], Y. Wang[1], [1]CNRS UMR6164, Polytech Nantes, Nantes, France, [2]IUT de Nantes, Carquefou, France

---

MO1D-4 Wideband Six-Port Receiver using Elliptical Microstrip-Slot Directional Couplers

M. Wei[1], Y. Chen[2], S. Qayyum[1], C. Tseng[2], R. Negra[1], [1]RWTH Aachen University, Aachen, Germany, [2]National Taiwan University of Science and Technology, Taipei, Taiwan

09:20

MO1B-4 Design of Fully Integrated Receiver Front-End for VSAT Applications

P. Wang[1], Y. Shen[1], M. Chou[1], Y. Chang[1,2], T. Wu[1], D. Chang[2], S. S. Hsu[1], [1]National Tsing Hua University, Hsinchu, Taiwan, [2]National Applied Research Laboratories, Hsinchu, Taiwan

---

MO1C-4 A New Form of Polynomial Model for Concurrent Dual-Band Digital Predistortion

C. Wang[1], W. Zhu[2], X. Zhu[1], [1]State Key Laboratory of Millimeter Waves, Nanjing, China, [2]School of Geography Science, Nanjing, China

MONDAY, 26 JANUARY 2015

RWW Session: MO2A	SiRF Session: MO2B	PAWR Session: MO2C	RWS Session: MO2D
RWW Distinguished Lecturers II Chair: Jeremy Muldavin, MIT Lincoln Laboratory Room: Gallery 2	**mmWave and Higher Frequency Applications** Chair: Herman Schumacher, Ulm University Co-Chair: Austin Chen, Skyworks Solutions Room: Grand Salon A	**High Efficiency RF Power Amplifiers** Chair: Robert Caverly, Villanova University Co-Chair: Art Morris , WiSpry Room: Gallery 1	**Emerging Technologies I** Chair: Lawrence Larson, Brown University Co-Chair: Medhi Shadaram, University of Texas at San Antonio Room: Grand Salon B

10:10

MO2A-1 An Overview of M-Health Medical Video Communications

Constantinos Pattichis, University of Cyprus

Co-Sponsored by IEEE EMBS

Abstract: Significant technological advances over the past decade have led M-health systems and services to a remarkable growth. It is anticipated that such systems and services will soon be established in standard clinical practice. M-health medical video communication systems progression has been primarily driven by associated advances in video coding and wireless networks technologies. This lecture reviews medical video communication systems. It highlights past approaches and focuses on current design trends and future challenges. It provides an insight to the most prevailing diagnostically driven concepts and the challenges associated with each system component.

MO2B-1 Review of Silicon-based Millimeter-wave Radio Frequeny Integrated Circuits (Invited)

H. Wang, National Taiwan University, Taipei, Taiwan

MO2C-1 Simplified Analysis and Design of Outphasing Transmitters Using Class-E Power Amplifiers (Invited)

R. A. Beltran[1], F. H. Raab[2], [1]Skyworks Solutions, Inc., Newbury Park, United States, [2]Green Mountain Radio Research Co., San Diego, United States

MO2D-1 3GPP ACLR and EVM Measurements for Millimeter-Wave Wireless Backhaul Applications at 60GHz

S. Maier, H. Schlesinger, G. Luz, D. Ferling, W. Kuebart, A. Pascht, Alcatel-Lucent Bell Labs Germany, Stuttgart, Germany

10:50

MO2B-2 A 122-150 GHz LNA with 30 dB Gain and 6.2 dB Noise Figure in SiGe BiCMOS Technology

R. Ben Yishay, D. Elad, E. Shumaker, IBM Haifa Research Lab, Haifa, Israel

MO2C-2 A Pulsed Load Modulation (PLM) Power Amplifier with 3-Level Envelope Delta-Sigma Modulation (EDSM)

Y. Song, R. Zhu, Y. E. Wang, University of California Los Angeles, Los Angeles, United States

MO2D-2 Remote Phase Synchronization for Satellite Network Systems

J. Xu[1], J. Long[2], D. Ye[1], J. Huangfu[1], C. Li[3], L. Ran[1], [1]Zhejiang University, Hangzhou, China, [2]University of California at San Diego, La Jolla, United States, [3]Texas Tech University, Lubbock, United States

11:10

MO2B-3 120 GHz Low Power, High Gain, Wideband Active Balun For Chip-to-Chip Communication

C. Lee, H. Lee, D. Kang, I. Song, H. Kim, S. Cho, J. Lee, I. Oh, C. Park, KAIST, Yuesong-gu, Republic of Korea

MO2C-3 A Full X-Band High-efficiency 12-Watt GaAs MMIC Power Amplifier with Harmonic Tuning

Q. Wu, B. Song, Y. Shih, X. Huang, J. Wu, RML Technology Co., Ltd, Chengdu, China

MO2D-3 Non-Contact Hand Interaction with Smart Phones Using the Wireless Power Transfer Features

C. Liu[1], C. Gu[2], C. Li[1], [1]Texas Tech University, Lubbock, United States, [2]Marvell Semiconductor Inc., Santa Clara, United States

11:30

MO2B-4 Electronic THz Transmissive Imaging System

W. Chen[1], C. Lai[1], T. Yan[1], C. Li[2], T. Chao[1], C. Kuo[1], [1]National Chiao-Tung University, Hsinchu City, Taiwan, [2]National Central University, Jhongli City, Taiwan

MO2C-4 A 400 W 2-Way Asymmetrical Doherty PA with 50% Efficiency Based on Second-Generation Airfast™ LDMOS Technology

S. Embar, L. Wang, J. Kim, C. Dragon, G. Tucker, Freescale Semiconductor Inc., Tempe, United States

MONDAY, 26 JANUARY 2015

RWS Session: MO3A	RWS-SiRF Joint Session: MO3B	PAWR Session: MO3C	RWS Session: MO3D
MM-Wave and THz	**Analysis and Arrays**	**CMOS RF Power Amplifier Technology**	**Emerging Technologies II**
Chair: Hasan Sharifi, HRL Labs Co-Chair: Rashaunda Henderson, University of Texas at Dallas	Chair: Jeremy Muldavin, MIT Lincoln Laboratory Co Chair: Tommy Ellis	Chair: Marc Franco, Qorvo Co-Chair: Murat Eron, Wireless Telecom Group	Chair: Khanna Rahul, Intel
Room: Gallery 2	Room: Grand Salon A	Room: Gallery 1	Room: Grand Salon B

13:30

MO3A-1 Microwave and Millimeter Wave Power Amplifiers: Technology, Applications, Benchmarks, and Future Trends (Invited)

J. J. Komiak, BAE Systems, Nashua, United States

MO3B-1 Front-End Non-Linear Distortion and Array Beamforming (Invited)

D. Rabinkin, W. Song, MIT Lincoln Laboratory, Lexington, United States

MO3C-1 CMOS High Bandwidth Envelope Tracking and Power Amplifiers for LTE Carrier Aggregation (Invited)

F. Balteanu, Skyworks Solutions, Inc., Irvine, United States

MO3D-1 Design of a Patch Antenna with Thermo-Electric Generator and Solar Cell for Hybrid Energy Harvesting

M. Virili[1,2], A. Georgiadis[2], A. Collado[2], P. Mezzanotte[1], L. Roselli[1], [1]University of Perugia, Perugia, Italy, [2]Centre Tecnològic de Telecomunicacions de Catalunya, Castelldefels, Spain

13:50

MO3D-2 Design of Efficient Rectifier for Low-Power Wireless Energy Harvesting at 2.45 GHz

T. Lee[1], P. Patil[1], C. Hu[1,2], M. Rajabi[1], S. Farsi[1], D. M. Schreurs[1], [1]KU Leuven, Heverlee, Belgium, [2]National Chiao Tung University, Hsinchu, Taiwan

14:10

MO3A-2 Power Synthesis at Low Frequencies in the THz Gap

J. Zhao[1], Z. Zhu[2], B. Zhang[1], D. Ye[1], C. Li[3], L. Ran[1], [1]Zhejiang Univercity, Hangzhou, China, [2]National Key Laboratory of Science and Technology on Space Microwave, Xian, China, [3]Texas Tech University, Lubbock, United States

MO3B-2 Continuous-Time Mode 2-D IIR Filter Enhanced Time-Delay Linear Aperture Arrays

A. Madanayake[1], C. Wijenayake[1], L. Belestotski[2], [1]The University of Akron, Akron, United States, [2]The University of Calgary, Calgary, Canada

MO3C-2 A 28 nm Standard CMOS Watt-Level Power Amplifier for LTE Applications

J. Fuhrmann[1,2], P. Oßmann[3], K. Dufrêne[1], H. Pretl[1], R. Weigel[2], [1]Friedrich-Alexander-University Erlangen-Nuremberg, Erlangen, Germany, [2]DMCE GmbH & Co. KG, Linz, Austria, [3]Johannes Kepler University, Linz, Austria

MO3D-3 Stability of Non-Foster Circuits for Broadband Impedance Matching of Electrically Small Antennas

A. M. Elfrgani, R. G. Rojas, The Ohio State University, Columbus, United States

14:30

MO3A-3 Evolution of DIG Integrated Platform for Millimeter-Wave Applications

M. A. Basha[1,2], A. Samir[2,1], R. Zaghloul[1], [1]Zewail City of Science and Technology, 6th of October City, Egypt, [2]Mansoura University, Mansoura, Egypt

MO3B-3 Intermittently Operating RF Frontend with 5ns Startup Time for 10Gbps Proximity Wireless Communication

N. Kitazawa, K. Kohira, H. Ishikuro, Keio University, Yokohama, Japan

MO3C-3 A Compact, High-gain Q-Band Stacked Power Amplifier in 45nm SOI CMOS With 19.2 dBm Psat and 19% PAE

W. Tai[1], D. S. Ricketts[2], [1]Carnegie Mellon University, Pittsburgh, United States, [2]North Carolina State University, Raleigh, United States

14:50

MO3A-4 A 30 GHz Impulse Radiator with On-Chip Antennas for High-Resolution 3D Imaging

P. Chen, A. Babakhani, Rice University, Houston, United States

MO3B-4 A New Multiple-Antenna-Port and Multiple-User-Port Antenna Tuner

F. Broyde, E. Clavelier, Excem, Maule, France

MO3C-4 Millimeter-wave Packaging on Alumina Board for E-band CMOS Power Amplifiers

Y. Zhang, D. Zhao, P. Reynaert, KU Leuven, Leuven, Belgium

MONDAY, 26 JANUARY 2015

Time: 14:20 – 16:10 — RWW STUDENT PAPER CONTEST
Room: Grand Salon CDE

RWW 2015 Student Paper Chairs will select finalists among the student paper submissions, from each conference (RWS, PAWR, BioWireleSS, and WiSNet, SiRF). During the poster presentation, held January 26, Monday afternoon 14:20 -16:10, judges will visit the student posters and grade the papers in the following five areas: novelty of the research, quality of the poster, quantity of information presented, preparedness of the presenter, and interest to the RWW community. The committee of judges will then select the first- and the second-place winners from each conference for a total of 8 winners. The awards will be announced and presented during the RWW Banquet held Tuesday night from 18:00-21:00. Please visit the student paper competition and support outstanding work by future researchers in industry and academia.

[MO3C-2] A 28 nm Standard CMOS Watt-Level Power Amplifier for LTE Applications
J. Fuhrmann[1,2], P. Oßmann[3], K. Dufrêne[1], H. Pretl[1], R. Weigel[2], [1]DMCE GmbH & Co. KG, Linz, Austria, [2]Friedrich-Alexander-University Erlangen-Nuremberg, Erlangen, Germany, [3]Johannes Kepler University, Linz, Austria

[MO3C-4] Millimeter-wave Packaging on Alumina Board for E-band CMOS Power Amplifiers
Y. Zhang, D. Zhao, P. Reynaert, KU Leuven, Leuven, Belgium

[MO4C-2] Characterization and Modeling of Pulse Drivers for Switch Mode Power Amplifier Measurements
N. Leder, T. I. Faseth, H. A. Ruotsalainen, H. Arthaber, Technische Universität Wien, Vienna, Austria

[WE1A-4] Low-Weight Wireless Sensor Network for Encounter Detection of Bats
M. Hierold[1], S. Ripperger[2], D. Josic[2], F. Mayer[2], R. Weigel[1], A. Koelpin[1], [1]University of Erlangen-Nuremberg, Erlangen, Germany, [2]Museum of Natural History, Berlin, Germany

[WE3D-5] Diode Detector Design for 61 GHz Substrate Integrated Waveguide Six-Port Radar Systems
S. Mann, S. Erhardt, S. Lindner, F. Lurz, S. Linz, F. Barbon, R. Weigel, A. Koelpin, University of Erlangen-Nuremberg, Erlangen, Germany

[WE4A-4] Underwater Interferometric Radar Sensor for Distance and Vibration Measurement
M. Sporer[1], F. Lurz[1], E. Schluecker[2], R. Weigel[1], A. Koelpin[1], [1]University of Erlangen-Nuremberg (Inst. Elec. Eng.), Erlangen, Germany, [2]University of Erlangen-Nuremberg (Inst. Proc. Tech. and Mach.), Erlangen, Germany

[WE3A-2] 100 GHz Reflectometer for Sensitivity Analysis of MEMS Sensors Comprising an Intermediate Frequency Six-port Receiver
S. Linz, F. Oesterle, A. Talai, S. Lindner, S. Mann, F. Barbon, R. Weigel, A. Koelpin, Friedrich-Alexander-University of Erlangen-Nuremberg, Erlangen, Germany

[TU1D-1] A Wearable System for Highly Selective L-glutamate Neurotransmitter Sensing
C. M. Nguyen[1], J. Mays[1], H. Cao[2], H. Allard[1], S. Rao[1], J. Chiao[1], [1]University of Texas at Arlington, Arlington, United States, [2]ETS Montreal, Montreal, Canada

[TU1D-3] A Low Power Wireless Sleep Apnea Detection System Based on Pyroelectric Sensor
I. Mahbub[1], M. Hasan[1], S. A. Pullano[2], F. Quaiyum[1], C. P. Stephens[3], S. K. Islam[1,3], A. S. Fiorillo[2], M. S. Gaylord[4], V. S. Lorch[4,3], N. Beitel[1,3], [1]University of Tennessee, Knoxville (ECE), Knoxville, United States, [2]University Magna Græcia, Catanzaro, Italy, [3]University of Tennessee, Knoxville (IBME), Knoxville, United States, [4]University of Tennessee, Knoxville (Mechanical,Aerosopace, BioMed), Knoxville, United States

[TU3D-1] Noncontact Heartbeat Detection using UWB Impulse Doppler Radar
L. Ren[1], Y. Koo[1], Y. Wang[2], A. E. Fathy[1], [1]University of Tennessee, Knoxville, Knoxville, United States, [2]Qorvo, Billerica, United States

[WE2D-2] A Permittivity Sensitive Phase-Locked Loop Based on a Silicon-Integrated Capacitive Sensor for Microwave Biosensing Applications
J. Nehring[1], M. Bartels[1], R. Weigel[1], D. Kissinger[1,2,3], [1]University of Erlangen-Nuremberg, Erlangen, Germany, [2]IHP GmbH, Frankfurt (Oder), Germany, [3]Technische Universität Berlin, Berlin, Germany

[MO3B-3] Intermittently Operating RF Frontend with 5ns Startup Time for 10Gbps Proximity Wireless Communication
N. Kitazawa, K. Kohira, H. Ishikuro, Keio University, Yokohama, Japan

[WE1C-2] An Integrated Reconfigurable Tuner in 45nm CMOS SOI Technology
A. Jou, C. Liu, S. Mohammadi, Purdue Univeristy, West Lafayette, United States

[WE2C-2] A 20GHz Class-C VCO Using Noise Sensitivity Mitigation Technique
K. Kimura, K. Okada, A. Matsuzawa, Tokyo Institute of Technology, OOkayama, Meguro-ku, Japan

[WE2C-4] Phase Noise Reduction in RF Oscillators utilizing Self-Injection Locked and Phase locked Loop
L. Zhang[1], A. K. Poddar[2], U. L. Rohde[2], A. S. Daryoush[1], [1]Drexel University, Philadelphia, United States, [2]Synergy Microwave, Paterson, United States

[MO2D-2] Remote Phase Synchronization for Satellite Network Systems
J. Xu[1], J. Long[2], D. Ye[1], J. Huangfu[1], C. Li[1], L. Ran[1], [1]Zhejiang University, Hangzhou, China, [2]University of California at San Diego, La Jolla, United States, [3]Texas Tech University, Lubbock, United States

[MO3A-4] A 30 GHz Impulse Radiator with On-Chip Antennas for High-Resolution 3D Imaging
P. Chen, A. Babakhani, Rice University, Houston, United States

[MO3D-3] Stability of Non-Foster Circuits for Broadband Impedance Matching of Electrically Small Antennas
A. M. Elfrgani, R. G. Rojas, The Ohio State University, Columbus, United States

[MO3D-1] Design of a Patch Antenna with Thermo-Electric Generator and Solar Cell for Hybrid Energy Harvesting
M. Virili[1,2], A. Georgiadis[2], A. Collado[2], P. Mezzanotte[1], L. Roselli[1], [1]University of Perugia, Perugia, Italy, [2]Centre Tecnològic de Telecomunicacions de Catalunya, Castelldefels, Spain

[MO4A-3] Curved Spiral Antennas for Freshwater Applications
R. A. Llamas[1,2], J. J. Niemeier[2], A. Kruger[1,2], [1]University of Iowa-ECE, Iowa City, United States, [2]University of Iowa-IIHR, Iowa City, United States

[MO4A-4] Fence Loaded Antenna Coupler for High-Band UWB with Steep Cutoff Characteristics
I. Saito, K. Kohira, H. Ishikuro, Keio University, Yokohama, Japan

[WE1B-5] A Class of Planar Multi-Band Wilkinson-Type Power Divider with Intrinsic Filtering Functionality
R. Loeches-Sanchez[1,2], D. Psychogiou[2], D. Peroulis[2], R. Gomez-Garcia[1], [1]University of Alcala, Alcala de Henares, Spain, [2]Purdue University , West Lafayette, United States

[WE4B-3] All-Analog Peak-to-Average Power Reduction using Constrained Clipping for OFDM Systems
M. Cho, J. S. Kenney, Georgia Institute of Technology, Atlanta, United States

[WE4B-4] An Experimental Evaluation on EVM Performance for 4-CSK(Color Shift Keying) using Visible Light with Multiple Full-color LEDs
H. Shimamoto, Y. Kozawa, Y. Umeda, Tokyo University of Science, Noda, Japan

Time: 14:00 – 16:20 — RWW MicroApps
Room: Grand Salon CDE

The Microwave/RF/Wireless Applications (MicroApps for short) Forum is a special session held within the exhibition area at RWW to enable vendors to conduct application-centric presentations (20 minutes in duration) that highlight their state-of-the-art products, technologies, services or solutions of interest to the RF/microwave and wireless community.

Company	Time Slot	Title
National Instruments	14:00-14:20	"Transceiver Module and Multi-element Phased Array Design with NI AWR Design Environment/Visual System Simulator (VSS) Software"
Sonnet	14:20-14:40	"The Unified-FFT Method Accelerating Wireless Designs"
Further MicroApps to be announced soon...		

MONDAY, 26 JANUARY 2015

RWS Session: MO4A	SiRF Session: MO4B	PAWR Session: MO4C	RWS Session: MO4D
Advanced Antenna Technology	**RF SOI Technologies and Applications**	**Power Amplifier Systems Concepts**	**Transceivers**
Chair: Nuno Borges Carvalho, Universidade de Aveiro	Chair: Paul Hurwitz, Tower Jazz Co-Chair: Mehmet Kaynak, IHP GmbH	Chair: Andrei Grebennikov, Microsemi Corporation Co-Chair: Almudena Suarez, University of Cantabria	Chair: Hiroshi Okazaki, NTT DoCoMo Inc.
Room: Grand Salon A	Room: Grand Salon B	Room: Gallery 1	Room: Gallery 2

15:40

MO4A-1 Design of Horizontally Polarized Ultra Wideband Slot Antennas for Wireless Applications

R. Kumar[1,2], R. Ram Krishna[2,1], [1]ARDE, Pune, India, [2]DIAT, Deemed University, Pune, India

MO4B-1 RFSOI Programmable Array of Capacitors (Invited)

M. Granger-Jones[1], J. Bendixen[3], J. Costa[2], M. Carroll[2], D. Kerr[2], C. Iversen[3], P. Mason[2], E. Spears[2], [1]Qorvo, San Jose, United States, [2]Qorvo, Greensboro, United States, [3]Qorvo, Aalborg, Denmark

MO4C-1 Challenges of Power Amplifier Design for Envelope Tracking Applications (Invited)

G. Collins[1], J. Wood[2], B. Woods[2], [1]MaxXentric Technologies, San Diego, United States, [2]Maxim Integrated, San Diego, United States

MO4D-1 SiGe BiCMOS Power Amplifier with a Switchable Output Matching Network for Efficiency Enhancement

Y. Lee, H. Li, J. Fu, National Central University, Jhongli City, Taiwan

16:00

MO4B-2 Improvements in SOI Technology for RF Switches (Invited)

M. Jaffe, A. Botula, J. Gambino, Z. He, A. Joseph, M. Abou-Khalil, R. Phelps, S. Shank, J. Slinkman, R. Wolf, J. Ellis-Monaghan, J. Gross, IBM Microelectronics Division, Essex Junction, United States

MO4D-2 A 9.99 mW Low-Noise Amplifier for 60 GHz WPAN System and 77 GHz Automobile Radar System in 90 nm CMOS

Y. Lin, C. Lee, C. Chen, National Chi Nan University, Puli, Taiwan

16:20

MO4A-3 Curved Spiral Antennas for Freshwater Applications

R. A. Llamas[1,2], J. J. Niemeier[2], A. Kruger[1,2], [1]University of Iowa-ECE, Iowa City, United States, [2]University of Iowa-IIHR, Iowa City, United States

MO4B-3 High Resistivity SOI Wafer Mapping for Mainstream RF System-on-Chip (Invited)

J. Raskin[1], E. Desbonnets[2], [1]Université Catholique de Louvain (UCL), Place du Levant[3], Belgium, [2]Soitec, Bernin, United States

MO4C-2 Characterization and Modeling of Pulse Drivers for Switch Mode Power Amplifier Measurements

N. Leder, T. I. Faseth, H. A. Ruotsalainen, H. Arthaber, Technische Universität Wien, Vienna, Austria

MO4D-3 30-GHz mHEMT Divide-by-Three Injection-Locked Frequency Divider With Marchand Balun

W. Chang[1], C. Meng[1], K. Tsung[1], G. Huang[2], [1]National Chiao Tung University, Hsinchu, Taiwan, [2]National Nano Device Laboratories, Hsinchu, Taiwan

16:40

MO4A-4 Fence Loaded Antenna Coupler for High-Band UWB with Steep Cutoff Characteristics

I. Saito, K. Kohira, H. Ishikuro, Keio University, Yokohama, Japan

MO4B-4 Comparison of Substrate Effects in Sapphire, Trap-Rich and High Resistivity Silicon Substrates for RF-SOI Applications

V. Sekar[1], C. Cheng[1], C. Zeng[1], A. Genc[2], T. Ranta[1], F. Rotella[1], R. Whatley[1], [1]Peregrine Semiconductor Corporation, San Diego, United States, [2]Entropic Communications, San Diego, United States

MO4C-3 The Impact of Channel Spacing on Memory Polynomial Models in Concurrent Dual-Band RF Power Amplification

K. N. Gebremicael[1], K. Morris[1], M. Beach[1], S. Wales[2], S. Ben Smida[1], M. Collett[1], [1]University of Bristol, Bristol, United Kingdom, [2]Chemring Technology Solutions, Hampshire, United Kingdom

MO4D-4 A GaN HEMT N-Path Filter with +17 dBm Jammer Tolerance

C. M. Thomas[1,2], L. E. Larson[3], [1]MaXentric Technologies, San Diego, United States, [2]University of California, San Diego, La Jolla, United States, [3]Brown University, Providence, United States

17:00

MO4A-5 FDTD Analysis of Platform Effect Reduction with Thin Film Ferrite

Z. Yao, Q. Xu, Y. E. Wang, University of California, Los Angeles, Los Angeles, United States

MO4B-5 Realisation of 40 GHz Conductor-backed Coplanar Waveguides and Meander Inductors on Surface-passivated High Resistivity Silicon

N. Hashim, A. Abuelgasim, K. de Groot, University of Southampton, Southampton, United Kingdom

MO4C-4 Bandwidth Reduction in Dynamic Load-modulated Power Amplifiers: Control and RF Signal Expansion, Efficiency and Linearity Trade-offs

K. Mimis, G. T. Watkins, Toshiba Research Europe Limited, Bristol, United Kingdom

TUESDAY, 27 JANUARY 2015

TU2 Plenary Session

Dr. Chris Van Hoof

imec/KULeuven

Time: 10:10-12:10
Room: Grand Salon AB

RWS Session: TU1B

Software Defined Radios and Cognitive Radios

Chair: Abbas Omar, University of Magdeburg

Room: Gallery 1

SiRF Session: TU1C

Topics in RF Modeling and Characterization Techniques

Chair: Hasan Sharifi, HRL Laboratories
Co-Chair: Monte Miller, Frescale

Room: Grand Salon B

BioWireleSS Session: TU1D

Wireless Remote Sensing of Biosignals

Chair: Mohamed Tofighi, Pennsylvania State University
Co-Chair: Aly Fathy, University of Tennessee

Room: Gallery 2

Title:
Wearable Wireless Sensor Technologies for Truly Personalized Medicine and Wellness

Abstract:
The healthcare arena is on a clear path towards preventative and personalized medicine. Semiconductor nanotechnologies are a key enabler to reaching this goal. Wearable sensors that combine ultra-low-power analog, digital, and radio circuitry and novel sensors are emerging. These wearables can measure progressively more parameters with greater accuracy, have become smaller in size and are more energy efficient-such that their continuous use becomes a practical reality. Apart from managing chronic illness, a further disruption in our healthcare will happen through the management of health where prediction and prevention will be essential enablers. Particularly in this domain, wearable and even disposable sensors that monitor whether you live a healthy life, that assess your stress levels, your pain, your emotions and so on, are examples of new tools that are moving out of the realm of science fiction and into everyday reality. This talk will describe such systems that achieve medically relevant information in a consumer form factor: wristbands, patches, headsets, smart textiles, smart contact lenses are promising wearable platforms that have the potential to create a multitude of killer apps - and these killer apps will be saving lives.

Plenary Speaker:
Dr. Chris Van Hoof, imec/KULeuven

Chris Van Hoof is Director of Wearable Healthcare at imec in Leuven, Belgium and Eindhoven, the Netherlands and imec Fellow. In the Wearable Healthcare program, imec and its industrial partners from across the value chain create and validate solutions at technology, component and application level. Chris Van Hoof has a track record of 20 years of initiating, executing and leading national and international contract R&D at imec. His work resulted in 3 startups (2 in the healthcare domain) and he delivered space qualified flight hardware to two cornerstone European Space Agency missions. After a PhD in Electrical Engineering (University of Leuven, 1992), Chris Van Hoof has held positions at imec at manager and director level in diverse technical fields (sensors and imagers, MEMS and autonomous microsystems, wireless sensors, body-area networks). He has published over 500 papers in journals and conference proceedings and given over 50 invited talks. Chris Van Hoof is also full professor at the University of Leuven (KULeuven).

08:00

TU1B-1 Receiver Cancellation of Radar in Radio

K. L. Tokuda[1], J. H. Kim[2], R. J. Baxley[1,2], J. S. Kenney[1], L. S. Cohen[3], [1]Georgia Institute of Technology, Atlanta, United States, [2]Georgia Tech Research Institute, Atlanta, United States, [3]Naval Research Laboratory, Washington, United States

TU1C-1 Tunable Filters and Antennas for 4G LTE Systems (Invited)

G. M. Rebeiz, C. H. Ko, Y. Cho, B. Avser, A. Alazemi, O. Gurbuz, University of California, San Diego, United States

TU1D-1 A Wearable System for Highly Selective L-glutamate Neurotransmitter Sensing

C. M. Nguyen[1], J. Mays[1], H. Cao[2], H. Allard[1], S. Rao[1], J. Chiao[1], [1]University of Texas at Arlington, Arlington, United States, [2]ETS Montreal, Montreal, Canada

8:20

TU1B-2 Double Quadrature Mixer for Adaptive Spur Cancellation in Ultra-Wideband Radios

S. Z. Mack[1,2], J. Wight[1,2], [1]Carleton University, Ottawa, Canada, [2]D-TA Systems, Ottawa, Canada

8:20

TU1D-2 Body-worn Fully-Passive Wireless Analog Sensors for Biopotential Measurement Through Load Modulation

S. Consul-Pacareu, D. Arellano, B. I. Morshed, The University of Memphis, Memphis, United States

08:40

TU1B-3 On the Double Threshold Energy Detection-Based Spectrum Sensing over kappa-mu Fading Channels

A. O. Ribas, U. S. Dias, University of Brasilia, Brasília, Brazil

TU1C-2 Multitone-FM Analysis of MEMS Varactor Phase Noise Contribution in VCOs

G. Kahmen[1], H. Schumacher[2], [1]Rohde & Schwarz GmbH, Munich, Germany, [2]Ulm University, Ulm, Germany

TU1D-3 A Low Power Wireless Sleep Apnea Detection System Based on Pyroelectric Sensor

I. Mahbub[1], M. Hasan[1], S. A. Pullano[2], F. Quaiyum[1], C. P. Stephens[3], S. K. Islam[1,3], A. S. Fiorillo[2], M. S. Gaylord[4], V. S. Lorch[4,3], N. Beitel[1,3], [1]University of Tennessee, Knoxville (ECE), Knoxville, United States, [2]University Magna Græcia, Catanzaro, Italy, [3]University of Tennessee, Knoxville (IBME), Knoxville, United States, [4]University of Tennessee, Knoxville (Mechanical,Aerosopace, BioMed), Knoxville, United States

09:00

TU1C-3 L-2L De-embedding Method with Double-T-type PAD Model for Millimeter-wave Amplifier Design

S. Kawai, K. K. Tokgoz, K. Okada, A. Matsuzawa, Tokyo Institute of Technology, Meguro-ku, Japan

TU1D-4 Feasibility of Patterned Vertical CNT for Dry Electrode Sensing of Physiological Parameters

M. Abu-Saude, S. Consul-Pacareu, B. I. Morshed, University of Memphis, Memphis, United States

09:20

TU1C-4 Cross-Line Characterization for Capacitive Cross Coupling in Differential Millimeter-Wave CMOS Amplifiers

K. K. Tokgoz, K. Lim, Y. Seo, S. Kawai, K. Okada, A. Matsuzawa, Tokyo Institute of Technology, Tokyo, Japan

TU1D-5 A Wireless Device to Monitor Pressure in Compression Bandages

N. Mehmood[1], A. Hariz[1], S. Templeton[2], N. H. Voelcker[3], [1]University of South Australia, Adelaide, Australia, [2]Royal District Nursing Service, Adelaide, Australia, [3]University of South Australia, Adelaide, Australia

TUESDAY, 27 JANUARY 2015

RWS Session: TU3A	RWS Session: TU3B	SiRF Session: TU3C	BioWireleSS Session: TU3D
Passive Antennas	**Propagation and Channel Modelling**	**Power Amplifier Applications**	**Remote Patient Monitoring and Energy Scavenging**
Chair: Jiang Zhu, Google[x]	Chair: Chenming Zhou, National Institute for Occupational Safety and Health	Chair: Julio Costa, Qorvo Co-Chair: Paul Hurwitz, Tower Jazz	Chair: Syed Kamrul Islam, University of Tennessee Co-Chair: Victor Lubecke, University of Hawaii
Room: Grand Salon A	Room: Gallery 1	Room: Grand Salon B	Room: Gallery 2

13:30

TU3A-1 A CPW-Fed Meandered-Shaped Monopole Antenna with Asymmetrical Ground Planes

D. Hsieh, J. Wu, Y. Cheng, C. Wang, National University of Tainan, Tainan, Taiwan

TU3B-1 Physics-based Ultra-Wideband Channel Modeling for Tunnel/Mining Environments

C. Zhou, National Institute for Occupational Safety and Health, Pittsburgh, United States

TU3C-1 A +18 dBm Broadband CMOS Power Amplifier RFIC with Distortion Cancellation

A. M. El-Gabaly[1,2], C. E. Saavedra[1], [1]Queen's University, Kingston, Canada, [2]Peraso Technologies Inc., Toronto, Canada

TU3D-1 Noncontact Heartbeat Detection Using UWB Impulse Doppler Radar

L. Ren[1], Y. Koo[1], Y. Wang[2], A. E. Fathy[1], [1]University of Tennessee, Knoxville, Knoxville, United States, [2]Qorvo, Billerica, United States

13:50

TU3A-2 WLAN Antenna Integrated in Indoor Ceiling Mounted Light System

L. Loizou, J. Buckley, B. O'Flynn, J. Barton, Tyndall National Institute, Cork, Ireland

TU3B-2 Millimeter-Wave Channel Sounding of Outdoor Ground Reflections

R. J. Weiler[1], M. Peter[1], W. Keusgen[1], A. Kortke[2], M. Wisotzki[1], [1]Fraunhofer Heinrich Hertz Institute, Berlin, Germany, [2]TU Berlin, Berlin, Germany

TU3C-2 A 1.8 to 2.4 GHz Stacked Power Amplifier Implemented in 0.25 μm CMOS SOS Technology

S. R. Helmi, H. Shan, S. Mohammadi, Purdue University, West Lafayette, United States

TU3D-2 Signal Processing Techniques for Vital Sign Monitoring Using Mobile Short Range Doppler Radar

A. Rahman, E. Yavari, X. Gao, V. M. Lubecke, O. Boric-Lubecke, University of Hawaii at Manoa, Honolulu, United States

14:10

TU3A-3 Dual-Band Pattern-Reconfigurable Yagi-Uda Antenna

N. Gagnon, Communications Research Centre Canada, Ottawa, Canada

TU3B-3 Spectrum Sensing over Nakagami-m/Gamma Composite Fading Channel with Noise Uncertainty

W. A. Silva, K. M. Mota, U. S. Dias, University of Brasilia, Brasilia, Brazil

TU3C-3 Channelized Active Noise Elimination (CANE) With Envelope Delta Sigma Modulation

R. Zhu, Y. Song, Y. E. Wang, University of California, Los Angeles, Los Angeles, United States

14:30

TU3A-4 Broadband RCS Reduction and Gain Enhancement Microstrip Antenna Using Ground Plane Slotted AMC Superstrate

J. Gao, J. Y. Zheng, Y. X. Cao, H. H. Yang, Q. W. Li, D. Zhang, Air Force Engineering University, Xi'an, China

TU3B-4 A Radiation Pattern Diversity Antenna Operating at the 2.4 GHz ISM Band

S. Dumanli, Toshiba Research Europe Limited, Bristol, United Kingdom

TU3C-4 A 60 GHz Highly Reliable Power Amplifier with 13 dBm Psat 15% Peak PAE in 65 nm CMOS Technology

B. Moret[1,2], N. Deltimple[1], E. Kerherve[1], A. Larie[1], B. Martineau[2], D. Belot[2], [1]University of Bordeaux, Talence, France, [2]STMicroelectronics, Crolles, France

TU3D-4 A Low-Input-Voltage Wireless Power Transfer for Biomedical Implants

H. Jiang[1], K. Bai[1], W. Zhu[1], D. Lan[1,2], J. Zhang[1], J. Wang[2], M. Shen[3], R. J. Fechter[4], M. Harrison[4], S. Roy[5], [1]San Francisco State University, San Francisco, United States, [2]University of South Florida, Tampa, United States, [3]Aalborg University, Aalborg, Denmark, [4]UC San Francisco-Surgery, San Francisco, United States, [5]UC San Francisco-Bioengineering and Therapeutic Sciences, San Francisco, United States

---

JOINT RWW BANQUET

Tuesday Evening, 27 January 2015 from 18:30-21:00
Room: Gallery 3B

Join your friends, co-workers and fellow researchers in an informal setting of lively discussion, dinner and wine. In addition, see the student paper award winners from the RWS, PAWR, WiSNet, BioWireless and SiRF receive their awards.

Exhibits/MicroApps/Demo

Industry Exhibits: Monday 26 January 13:00 - 17:30 and Tuesday 27 January 10:00 - 17:00
Room: Grand Salons CDE
MicroApps Talks: Tuesday 27 January 14:00 - 16:20
Room: Grand Salons CDE
Demo Session: Tuesday 27 January 15:00 - 17:00
Room: Grand Salons CDE

TUESDAY, 27 JANUARY 2015

Interactive Poster Session: Power Amplifiers 14:55-16:55

TU3P Advances in RF Power Amplifiers

Chair: Yupeng Jia, National Instruments
Room: Grand Salon CDE

TU3P-1 On the use of Frequency Transformations in the Design of Broad-band and Concurrent Multi-band Power Amplifiers
N. Nallam[1], S. Chatterjee[2], [1]IIT Guwahati, Guwahati, India, [2]IIT Delhi, New Delhi, India

TU3P-2 Envelope Tracking RF Power Amplifier with Lead-Lag Modulator
G. T. Watkins, K. Mimis, Toshiba Research Europe Limited, Bristol, United Kingdom

TU3P-3 A 3.6 GHz Linear High Efficiency Doherty Amplifier with 40 dBm Saturated Output Power using GaN on SiC HEMT Devices
B. Baker[1,2], R. L. Campbell[2], [1]Qorvo, Hillsboro, United States, [2]Portland State University, Portland, United States

TU3P-4 On the Estimation of Power Amplifier Efficiency for Modulated Signals
M. Vejdaniamiri, M. Helaoui, F. Ghannouchi, University of Calgary, Calgary, Canada

TU3P-5 Bi-level Quadrature-modulation Low-pass EPWM transmitter Using Half Side of Tri-level $\Delta\Sigma$ Modulator
T. Noda, W. Someya, Y. Iikura, Y. Umeda, Y. Kozawa, Tokyo University of Science, Noda, Japan

TU3P-6 RF Power Amplifier Behavioral Modeling Using Wavelet Multiresolution
C. Mateo-Pérez, P. L. Carro, P. García-Dúcar, J. de Mingo, University of Zaragoza, Zaragoza, Spain

TU3P-7 Designing Power Amplifiers for Spectral Compliance Using Spectral Mask Load-Pull Measurements
M. Fellows, J. Barlow, C. Baylis, J. Barkate, R. J. Marks II, Baylor University, Waco, United States

TU3P-8 Over 65% PAE GaN Voltage-Mode Class D Power Amplifier for 465 MHz Operation Using Bootstrap Drive
H. Nakamizo[1,2], K. Mukai[1], S. Shinjo[1], H. Gheidi[2], P. Asbeck[2], [1]Mitsubishi Electric Corporation, Kamakura, Japan, [2]University of California, San Diego, La Jolla, United States

TU3P-9 GaN-on-Si Transformer-Coupled Class D Power Amplifier
M. R. Hasin[1], J. N. Kitchen[1], B. Ardouin[2], [1]Arizona State University, Tempe, United States, [2]XMOD Technologies, Bordeaux, France

TU3P-10 Study of the Impedance Transformation Ratio of Microwave Rectifier for Outphasing Power Recycling Application
D. Wang, J. Guan, R. Negra, RWTH Aachen University, Aachen, Germany

Exterior View of the Omni San Diego Hotel
Courtesy: Omni Hotel, San Diego

TUESDAY, 27 JANUARY 2015

Demo Track Presentations
Tuesday, 15:00- 17:00
Room: Grand Salon CDE

In its fourth year of RWW, there will be a demo session where presenters bring in demonstrations of their latest wireless experiments for a hands-on interactive forum. Come, see and feel how people are tackling real-world problems to address the next wireless innovation!

1. Noninvasive Continuous Mobile Blood Pressure Monitoring using Novel PPG Optical Sensor
Vahram Mouradian, Armen Poghosyan, and Levon Hovhannisyan, Sensogram Technologies Inc., USA
We are presenting a novel PPG optical sensor and methodology which have been integrated into a prototype standalone device ensuring for the first time the noninvasive, continuous, wearable, remote and mobile monitoring of blood pressure and other human vital signs, such as heart rate, oxygen saturation, respiration rate, etc. This small device allows the user to read, store, process and transmit all the measurements to a remote location.

2. Non-contact Hand Interaction with Smart Phones using the Wireless Power Transfer Features
Chenhui Liu, Changzhan Gu, and Changzhi Li, Department of Electrical and Computer Engineering, Texas Tech University & Marvell Semiconductor Inc., USA
We will demonstrate the non-contact interaction with the wireless power transfer coil inside the smart phone. The interaction between human and smart phones gradually changed from button pressing to screen touching in the past decade. Lately, as a trend of wireless application, wireless charging is growing up as a competitive feature for smart phones. The basic idea of the wireless charging is the wireless electromagnetic coupling between inductive coils, which means that we can also perform mutual coupling between our hand and the coil, so as to control smart phones without contact. This technique illustrates a system that configures an oscillator using the wireless charging coil as part of the resonant tank, and the resonant frequency of the oscillator will change as the impedance of the coil will change due to the mutual coupling between hand and the coil. This enables us to perform non-contact interaction with smart phones with little extra hardware expense.

3. Real-time Jammer Suppression Using Evanescent-mode Cavity Filters
Mohammad Abu Khater, Dimitra Psychogiou, and Dimitrios Peroulis, Adaptive Radio Electronics and Sensors (ARES) Group, Dpt. of Electrical and Computer Engineering, Purdue University, USA
With the ever increasing usage of the frequency radio spectrum, the performance of RF transceivers is severely degraded by radio frequency interference that is often created by adjacent electronic devices or coexisting communication carriers. In this demo, a real-time monitoring system that is able to: (a) identify and (b) suppress jamming signals is demonstrated for the first time. The jammer identification concept is based on a closed loop system that consists of a tunable evanescent-mode band-stop filter (BSF) followed by a power detector.

4. Perfect Wireless Power Receiving Surface
Zheda Chen, Rong Wang, Jiaqi Zhao, Dexin Ye, and Lixin Ran, Zhejiang University, China
The surface is an artificially synthesized perfectly matched layer (PML) embedded with Schottky rectifying diodes to harvest wireless energy existing in natural environment. The surface is well designed to allow maximum receiving and recycling of ubiquitous radio power from ambient-radiation sources and provide DC power for low-power electronics. Compared with conventional rectenna system, the surface can achieve a nearly perfect absorption of the ambient wireless energy with a large receiving area. With its simple structure, such artificial surface can be specially designed and tailored to maximize wireless energy absorption under different environment. It can also be economically produced in the forms of soft substrates and/or textures which can then be used in clothes, tents or other daily items for energy harvesting. The surface can also be used in other renewable energy fields such as Solar Power Satellite System (SPSS).

5. Real-time PreDistortion and Envelope Tracking for High Efficiency Power Amplifiers
Jonmei J. Yan, Paul Theilmann, Donald F. Kimball, and Toshifumi Nakatanii, MaXentric Technologies, LLC., USA
We will present a live demonstration of envelope tracking with real-time digital pre-distortion (DPD) for high efficiency and high linearity power amplifiers for micro-basestation applications. This work is motivated by today's need for high spectral efficiency in the crowded frequency spectrum allocations, leading to signals with high peak to average ratios (PAPRs). Unlike conventional power amplifiers, with the use of real-time pre-distortion, envelope tracking power amplifiers can achieve high efficiency and high linearity simultaneously. In envelope tracking, the power supplied to the RF power amplifier (RFPA) varies as a function of the envelope of the RF signal, minimizing the power consumption and increasing its efficiency by keeping the RFPA close to saturation most of the time.

6. Low Power 24 GHz Radar System for Occupancy Monitoring
Fabian Lurz, Sebastian Mann, Sarah Linz, Stefan Lindner, Robert Weigel, and Alexander Koelpin, Institute for Electronics Engineering, University of Erlangen-Nuremberg, Germany
We will demonstrate a low-power 24 GHz continuous wave (CW) prototype system for occupancy monitoring and presence detection. It is based on a minimalistic hardware approach and is able to wirelessly sense human respiratory signals so that even non-moving persons can be detected. By intermittently measuring, the average power consumption can be significantly reduced down to e.g. 0.2mW for 20 measurements per second. Experiments verify that, due to the short wavelength, the single channel receiver limitations can be neglected when only a detection of human presence but no evaluation of the breathing frequency is necessary. For the demo track session we propose a functional system demonstrator with a live MATLAB graphical user interface (GUI). Additional insight will be given into the internal processes of the low-power system concept by showing the baseband voltage and duty cycles of the single components on a digital storage oscilloscope (DSO) while simultaneously monitoring the power consumption on a precision DC analyzer.

7. Sub-THz Low-power and High-speed OOK Transmitter and Receiver for uncompressed HD video streaming
Hea Jin Lee, Chong Hyun Yoon, Joong Geun Lee, Chae Jun Lee, Dong Min Kang, In Sang Song, Sung Jun Cho, Hong Yi Kim, Inn Yeol Oh, and Chul Soon Park, Department of Electrical engineering, Korea Advanced Institute of Science and Technology (KAIST), Korea
Through researches, we designed low-power high-speed OOK transmitter and receiver using sub-THz carrier frequency and will demonstrate wireless streaming of large amount of data having a data rate of 3Gbps at this demo track. High speed characteristic of sub-THz enables the system to use the uncompressed data for transmitting and receiving so that the system architecture and its cost is reduced.

8. Compact High Resolution Radar at 80 GHz and 140 GHz
Nils Pohl, Sven Thomas, Simon Kueppers, and Timo Jaeschke, Institute of Integrated Systems, Ruhr-University Bochum, Germany
We will demonstrate an FMCW radar sensor with two different front-ends, operating at 80GHz and 140GHz. They achieve an ultra-wide bandwidth of 25GHz and 48GHz, respectively. The mmWave front-ends of the sensor are realized as a custom SiGe MMIC and embedded in a compact sensor board. For control and raw data transmission a USB interface is used. The signal processing and visualization can be done using a standard computer, e.g. with MATLAB. The sensor will be demonstrated with both front-ends as a live demo during the RWW demo session.

TUESDAY, 27 JANUARY 2015

RWS-SiRF Joint Session: TU5A	RWS Session: TU5B	RWS Session: TU5C	BioWireleSS Session: TU5D
RF & Internet of Things	**Late News**	**High Speed II**	**Wireless BAN and Medical Imaging**
Chair: Jeremy Muldavin, MIT Lincoln Laboratory Co-Chair: Karen Gettings, MIT Lincoln Laboratory	Chair: Sergio Pacheco, Freescale	Chair: Debabani Choudhury, Intel	Chair: Changzhi Li , Texas Tech University Co-Chair: Arnaud Pothier, XLIM
Room: Grand Salon A	Room: Gallery 1	Room: Grand Salon B	Room: Gallery 2

16:00

TU5A-1 Redefining the Leading Edge: A Silicon RF Perspective (Invited)

P. Colestock, Global Foundries, San Diego, United States

TU5B-1 A Pseudorandom Clocking Scheme for a CMOS N-path Bandpass Filter with 10-to-15 dB Spurious Leakage Improvement

C. Thomas[1], W. Leung[2], L. E. Larson[3], [1]University of California, San Diego, La Jolla, United States, [2]Qualcomm Inc., San Diego, United States, [3]Brown University, Providence, United States

TU5C-1 Iterative Receiver for Millimeter-Wave OFDM Systems: Evaluation of High Doppler Shift by Dynamic Channel Model

Y. Chang, M. Furukawa, H. Suzuki, K. Fukawa, Tokyo Institute of Technology, Tokyo, Japan

TU5D-1 Fiber Antenna for Wireless Body Area Networks

T. Nikoubin[1], M. Garipally[1], T. Nguyen[2], Z. Wang[2], M. Saed[1], C. Li[1], [1]Texas Tech University, Lubbock, United States, [2]University of Texas, Austin, United States

16:20

TU5B-2 Performance of Non-Coherent FSK Virtual MISO Systems in Correlated Rayleigh Fading

M. Hussain, S. Hassan, National University of Sciences & Technology, Islamabad, Pakistan

TU5C-2 Development of a very Low-cost Down Converter for the IEEE802.11ad Wireless Network Appliance Test

K. Fujiwara[1], N. Shibagaki[2], T. Kobayashi[1], H. Hanyu[1], [1]Tokyo Metropolitan Industrial Technology Research Institute, Koto-ku, Japan, [2]Hitachi, Ltd., Information & Telecommunication Systems Company, Kawasaki, Japan

TU5D-2 Radiation Pattern Steering for On-body Gateways at the 2.4 GHz ISM Band

S. Dumanli, Toshiba Research Europe Limited, Bristol, United Kingdom

16:40

TU5A-2 RF and Microwave Technology Challenges for Internet-of-Things Applications (Invited)

L. E. Larson, Brown University, Providence, United States

TU5B-3 A Low Power 24 GHz Radar System for Occupancy Monitoring

F. Lurz, S. Mann, S. Linz, S. Lindner, F. Barbon, R. Weigel, A. Koelpin, University of Erlangen-Nuremberg, Erlangen, Germany

TU5C-3 Performance Evaluation of LTE-Advanced Downlink Adopting Higher Order Modulation in Small Cells

T. Ohseki, T. Yamamoto, Y. Suegara, KDDI R&D Laboratories, Inc., Fujiminoshi, Japan

TU5D-3 Dual Thermal Time Constant Electrothermal Modeling of PIN Diode Protection Circuits

R. H. Caverly, Villanova University, Villanova, United States

17:00

TU5C-4 Evaluation of Information Leak by Robustness Evaluation of Countermeasure to Disguised CSI in PLNC Considering Physical Layer Security

K. Matsumoto[1], O. Takyu[1], T. Fujii[2], T. Ohtsuki[3], F. Sasamori[1], S. Handa[1], [1]Shinshu University, Nagano, Japan, [2]The University of Electro-Communications, Chofu, Japan, [3]Keio University, Yokohama, Japan

TU5D-4 Reconfigurable Analog-to-Digital Converter for Implantable Bioimpedance Monitoring

T. C. Randall, I. Mahbub, S. K. Islam, University of Tennessee, Knoxville, United States

WEDNESDAY, 28 JANUARY 2015

WiSNet Session: WE1A	RWW Session: WE1B	SiRF Session: WE1C	BioWireleSS Session: WE1D
Insight in Sensor Networks and System Design Chair: Rahul Khanna, Intel Co-Chair: Andreas Stelzer, Johannes Kepler University, Linz Room: Gallery 1	**Passive Components and Packaging I** Chair: Hualiang Zhang, University of North Texas Co-Chair: Roberto Gomez-Garcia, University of Alcala Room: Grand Salon A	**Tunable and Reconfigurable Technologies** Chair: J.P. Raskin, Université catholique de Louvain (UCL) Co-Chair:Monte Miller, Freescale Room: Grand Salon B	**Micro Biosensing** Chair: Dietmar Kissinger , IHP GmbH Co-Chair: JC Chiao, University of Texas Arlington Room: Gallery 2

08:00

WE1A-1 Review of the Present Technologies Concurrently Contributing to the Implementation of the Internet of Things (IoT) Paradigm: RFID, Green Electronics, WPT and Energy Harvesting (Invited) *L. Roselli[1], C. Mariotti[1], P. Mezzanotte[1], F. Alimenti[1], G. Orecchini[1], M. Virili[1], N. B. Carvalho[2], [1]University of Perugia, Perugia, Italy, [2]University of Aveiro, Aveiro, United States*	**WE1B-1 Miniaturized Via-less Ultra-Wideband Bandpass Filter Based on CRLH-TL Unit Cell** *A. O. Alburaikan, M. Aqeeli, X. Huang, Z. Hu, The University of Manchester, Manchester, United Kingdom*	**WE1C-1 Reconfigurable Solutions for Mobile Device RF Front-ends (Invited)** *A. Morris, wiSpry, San Diego, United States*	**WE1D-1 Why using High Frequency Dielectric Spectroscopy for Biological Analytics? (Invited)** *M. Poupot[1,2], D. Dubuc[2,3], F. Artis[1,3], K. Grenier[2,3], J. Fournie[1,2] [1]CRCT , Av. Hubert Curien, France, [2]Univ. Toulouse [3], Toulouse, France, [3]CNRS, Toulouse, France*

08:20

WE1A-2 Combined Localization and Data Transmissionin Energy-Constrained Wireless Sensor Networks *T. Nowak[1], A. Koelpin[2], F. Dressler[3], M. Hartmann[1], L. Patino[1], J. Thielecke[1], [1]University of Erlangen-Nürnberg-Inst. Info. Tech., Erlangen, Germany, [2]University of Erlangen-Nürnberg-Inst. Elec. Eng., Erlangen, Germany, [3] University of Paderborn, Paderborn, Germany*	**WE1B-2 Dual-Band Negative Group Delay Circuit Using Defected Microstrip Structure** *G. Chaudhary[1], P. Kim[1], J. Jeong[1], Y. Jeong[1], J. Lim[2], [1]Chonbuk National University, Jeonju-si, Republic of Korea, [2]Soonchunhyang University, Asan, Republic of Korea*		

08:40

WE1A-3 Wireless Integrated Sensor Nodes for Indoor Monitoring and Localization (Invited) *D. Kissinger[1,2], A. Schwarzmeier[3], F. Grimminger[4], J. Mena-Carrillo[4], W. Weber[4], G. Hofer[5], G. Fischer[3], R. Weigel[3], [1]IHP, Frankfurt (Oder), Germany, [2]Technische Universität Berlin, Berlin, Germany, [3]FAU Erlangen-Nuremberg, Erlangen, Germany, [4]Infineon Technologies, Neubiberg, Germany, [5]Infineon Technologies Austria, Graz, Austria*	**WE1B-3 A high power Ka-band SPST switch MMIC using 0.25 um GaN on SiC** *S. Kaleem[1], J. Kuhn[2], R. Quay[2], M. Hein[1], [1]Ilmenau University of Technology, Ilmenau, Germany, [2]Fraunhofer Society for the Advancement of Applied Research, Freiburg, Germany*	**WE1C-2 An Integrated Reconfigurable Tuner in 45nm CMOS SOI Technology** *A. Jou, C. Liu, S. Mohammadi, Purdue Univeristy, West Lafayette, United States*	**WE1D-2 Broadband Dielectric Characterization of CHO-K1 Cells Using Miniaturized Transmission-Line Sensor** *N. Meyne[1], G. Fuge[2], S. Hemanth[3], H. K. Trieu[3], A. Zeng[2], A. F. Jacob[1], [1]Technische Universität Hamburg-Harburg-Inst. Hochfreq., Hamburg, Germany, [2]Technische Universität Hamburg-Harburg-Inst. Bioproz. und Biosys., Hamburg, Germany, [3]Technische Universität Hamburg-Harburg-Inst. Mikrosystem., Hamburg, Germany*

09:00

WE1A-4 Low-Weight Wireless Sensor Network for Encounter Detection of Bats *M. Hierold[1], S. Ripperger[2], D. Josic[2], F. Mayer[1], R. Weigel[1], A. Koelpin[1], [1]University of Erlangen-Nuremberg, Erlangen, Germany, [2]Museum of Natural History, Berlin, Germany*	**WE1B-4 High Frequency-Selectivity Impedance Transformer** *P. Kim[1], G. Chaudhary[1], J. Park[1], Y. Jeong[1], J. Lim[2], [1]Chonbuk National University, Jeonju, Republic of Korea, [2]Soonchunhyang University, Asan, Republic of Korea*	**WE1C-3 Ferroelectric MIM Capacitors for Compact High Tunable Filters** *R. De Paolis[1], S. Payan[2], M. Maglione[2], G. Guegan[3], F. Coccetti[1], [1]CNRS, Toulouse, France, [2]CNRS, Bordeaux, France, [3]ST-Microelectronics, Tours, France*	**WE1D-3 A Microwave Sensor Dedicated to Dielectric Spectroscopy of Nanoliter Volumes of Liquid Medium and Flowing Particles** *A. Landoulsi, C. Dalmay, A. Bessaudou, P. Blondy, A. Pothier, XLIM, Limoges, France*

09:20

WE1A-5 Ad-Hoc Multilevel Wireless Sensor Networks for Distributed Microclimatic Diffused Monitoring in Precision Agriculture *A. Rodriguez de la Concepcion, R. Stefanelli, D. Trinchero, iXem Labs - Politecnico di Torino, Torino, Italy*	**WE1B-5 A Class of Planar Multi-Band Wilkinson-Type Power Divider with Intrinsic Filtering Functionality** *R. Loeches-Sanchez[1,2], D. Psychogiou[2], D. Peroulis[2], R. Gomez-Garcia[1], [1]University of Alcala, Alcala de Henares, Spain, [2]Purdue University , West Lafayette, United States*	**WE1C-4 10.6 THz Figure-of-Merit Phase-change RF Switches with Embedded Micro-heater** *J. Moon, H. Seo, D. Le, H. Fung, A. Schmitz, T. Oh, S. Kim, K. Son, B. Yang, HRL Laboratories, Malibu, United States*	**WE1D-4 Sub-microliter Microwave Dielectric Spectroscopy for Identification and Quantification of Carbohydrates in Aqueous Solution** *F. Artis[1,2], D. Dubuc[1], J. Fournie[2], M. Poupot[2], K. Grenier[1], [1]LAAS-CNRS and Toulouse Univ., Toulouse, France, [2]CRCT, Toulouse, France*

WEDNESDAY, 28 JANUARY 2015

WisNet Session: WE2A	RWW Session: WE2B	SiRF Session: WE2C	BioWireleSS Session: WE2D
Advanced Localization and Sensing Technologies Chair: Luca Roselli, University of Perugia Co-Chair: Holger Maune, University of Darmstadt Room: Gallery 1	**Passive Components and Packaging II** Chair: Dariush Mirshekar-Syahkal, University of Essex Co-Chair: Rashaunda Henderson, University of Texas Dallas Room: Grand Salon A	**SiRF Circuits and Applications - 2** Chair: Chiennan Kuo, National Chiao Tung University Co-Chair: Austin Chen, Skyworks Solutions Room: Grand Salon B	**Microwaves Interaction with Biological Materials** Chair: JC Chiao, University of Texas Arlington Co-Chair: Pinshan Wang, Clemson University Room: Gallery 2

10:10

WE2A-1 Robust Localization of Passive UHF RFID Tag Arrays Based on Phase-Difference-of-Arrival Evaluation M. Scherhäufl[1], M. Pichler[1], A. Stelzer[2], [1]Linz Center of Mechatronics GmbH, Linz, Austria, [2]Johannes Kepler University, Linz, Austria	**WE2B-1 Varactor Tuned Ring Resonator Filter With Wide Tunable Bandwidth** C. Kim[1], K. Chang[2], X. Liu[1], [1]University of California Davis, Davis, United States, [2]College Station, United States	**WE2C-1 Low Power and High Speed OOK Modulator for Wireless Inter-Chip Communications** H. Lee, C. Yoon, J. Lee, C. Lee, D. Kang, I. Song, S. Cho, H. Kim, I. Oh, C. Park, KAIST, 291 Daehak-ro, Yuseong-gu, Daejeon, Republic of Korea	**WE2D-1 When Dielectric Spectroscopy Meets THz Spectroscopy; The Tale of Two Estranged Brothers (Invited)** Y. Feldman[1], P. Ben Ishai[1,2], [1]The Hebrew University of Jerusalem, Jerusalem, Israel, [2]Neteera, Jerusalem, Israel

10:30

WE2A-2 Experimental Evaluation of A Pairwise Broadcast Synchronization in A Low-Power Cyber-Physical System U. Ghoshdastider, R. Viga, M. Kraft, University of Duisburg-Essen, Duisburg, Germany	**WE2B-2 Small Low-Pass Filter Using Slotted-Ground-Plane Resonator** J. Wu, D. Hsieh, Y. Cheng, W. Wang, C. Wang, National University of Tainan, Tainan, Taiwan	**WE2C-2 A 20GHz Class-C VCO Using Noise Sensitivity Mitigation Technique** K. Kimura, K. Okada, A. Matsuzawa, Tokyo Institute of Technology, Okayama, Meguro-ku, Japan	

10:50

WE2A-3 DMA-driven Control Method for Low Power Sensor Node T. Enami, K. Kawakami, H. Yamazaki, Fujitsu Laboratories Ltd., Kawasaki, Japan		**WE2C-3 Radio-Frequency Flexible Transistors on Cellulose Nanofibrillated Fiber (CNF) Substrates** J. Seo[1], T. Chang[1], R. Sabo[2], Z. Cai[2], S. Gong[3], Z. Ma[1], [1]University of Wisconsin-Madison, Madison, United States, [2]U.S. Department of Agriculture (USDA), Madison, United States, [3]University of Wisconsin-Madison, Madison, United States	**WE2D-2 A Permittivity Sensitive Phase-Locked Loop Based on a Silicon-Integrated Capacitive Sensor for Microwave Biosensing Applications** J. Nehring[1], M. Bartels[1], R. Weigel[1], D. Kissinger[1,2,3], [1]University of Erlangen-Nuremberg, Erlangen, Germany, [2]IHP GmbH, Frankfurt (Oder), Germany, [3]Technische Universität Berlin, Berlin, Germany

11:10

WE2A-4 Wireless Sensors for Stratified Soil Microwave Scanning Applied to Precision Quality Agriculture E. Pievanelli, D. Trinchero, A. Rodriguez de la Concepcion, R. Stefanelli, iXem Labs - Politecnico di Torino, Torino, Italy	**WE2B-4 Sharp-Rejection Highpass and Dual-Band Bandpass Planar Filters with Multi-Transmission-Zero-Generation Transversal Cell** R. Loeches-Sanchez[1,2], D. Psychogiou[2], D. Peroulis[2], R. Gomez-Garcia[1], [1]University of Alcala, Alcala de Henares, Spain, [2]Purdue University, West Lafayette, United States	**WE2C-4 Phase Noise Reduction in RF Oscillators utilizing Self-Injection Locked and Phase locked Loop** L. Zhang[1], A. K. Poddar[2], U. L. Rohde[2], A. S. Daryoush[1], [1]Drexel University, Philadelphia, United States, [2]Synergy Microwave, Paterson, United States	**WE2D-3 Non-contact Measurement of Complex Permittivity Based on Coupled Magnetic and Electric Resonances** J. Dong[1], F. Shen[1], J. Huangfu[1], S. Qiao[2], C. Li[3], L. Ran[1], [1]Zhejiang University, Hangzhou, China, [2]Zhejiang University City College, Hangzhou, China, [3]Texas Tech University, Lubbock, United States

11:30

WE2A-5 Sensor Network with Energy Efficient and Low-cost Gas Sensor Nodes for the Detection of Hazardous Substances in the Event of a Disaster S. Rademacher, K. Schmitt, M. Mengers, J. Wöllenstein, Fraunhofer Institute for Physical Measurement Techniques IPM, Freiburg, Germany	**WE2B-5 A Fourth Order Tunable Capacitor Coupled Microstrip Resonator Band Pass Filter** S. Hao, Q. J. Gu, University of California, Davis, Davis, United States		**WE2D-4 Design and Evaluation of Electrode for Dielectrophoretic Characterization of Blood Cells** M. Eguchi[1], F. Kuroki[2], H. Imasato[1], T. Yamakawa[1], [1]Fuzzy Logic Systems Institute, Kitakyushu, Japan, [2]Kure National College of Technology, Kure, Japan

WEDNESDAY, 28 JANUARY 2015

Joint RWW Interactive Poster Session
12:55-14:30

WE3P: Transceivers and Front-end Technologies SOC and SiP

Chair: Yupeng Jia, National Instruments
Room: Grand Salon CDE

WE3P-1 A Low Power and High Conversion Gain 77~81 GHz Double-Balanced Up-Conversion Mixer with Excellent LO-RF Isolation in 90 nm CMOS
Y. Lin, R. Liu, C. Wang, C. Chen, National Chi Nan University, Puli, Taiwan, National Chi Nan University, Puli, Taiwan

WE3P-2 Accelerating Software Radio on ARM: Adding NEON Support to VOLK
N. E. West[2,1], D. J. Geiger[1], G. M. Scheets[2], [1]U.S. Naval Research Laboratory, Washington, United States, [2]Oklahoma State University, Stillwater, United States

WE3P-3 Series-Cascaded Absorptive Notch-Filters for 4G-LTE Radios
D. Psychogiou, R. Mao, D. Peroulis, Purdue University, West Lafayette, United States

WE3P-4 Two Half-Lambda Dipole Array Coplanar Feed Wideband PCB Antenna
Q. W. Pan, Manukau Institute of Technology, Manukau, New Zealand

WE3P-5 On Coupled-Resonator Filters with Tunable Bandwidth
M. K. Wohler, A. Jaschke, M. Schühler, Fraunhofer Institute for Integrated Circuits, Erlangen, Germany

WE3P-6 Performance Comparison of Raised Cosine Shaped and Rectangular Pulsed Signals in E-Band Communication Systems
F. Boes[1], J. Antes[1], D. Meier[1], T. Messinger[1], D. Müller[1,2], R. Henneberger[3], A. Tessmann[4], I. Kallfass[1], 1University of Stuttgart, Stuttgart, Germany, [2]Karlsruher Institute of Technology, Karlsruhe, Germany, [3]RPG Radiometer Physics GmbH, Meckenheim, Germany, [4]Frauenhofer IAF, Freiburg, United States

WE3P-7 Resonant Characteristics of Metal Rod Resonator Supported by PEEK Material at 60 GHz
M. Teramoto, F. Kuroki, National Institute of Technology, Kure College, Kure, Japan

WE3P-8 A Capacitively-loaded Loop Antenna for UHF Near-field RFID Reader Applications
M. Dhaouadi[1], M. Mabrouk[1], A. Ghazel[1], T. Phu Vuong[2], A. Coelho[2], [1]Grescom SUPCOM, Ariana, Tunisia, [2]Grenoble INP - Minatec , Grenoble, France

WE3P-10 Gold Nanorod Array Structured Silicon Nitride Films for Reliable RF MEMS Capacitive Switches
L. Michalas[1], S. Xavier[2], M. Koutsoureli[1], O. El Jouaidi[2], S. Bansropun[1], G. Papaioannou[2], A. Ziaei[2], [1]University of Athens, Athens, Greece, [2]Thales, Paris, France

WE3P-11 Noninvasive Continuous Mobile Blood Pressure Monitoring using Novel PPG Optical Sensor
V. Mouradian, A. Poghosyan, L. Hovhannisyan, Sensogram Technologies Inc., Plano, United States

WE3P-12 Miniaturized 60 GHz Triangular CMOS Antenna-on-Chip using Asymmetric Artificial Magnetic Conductor
A. Barakat[1], A. Allam[1], H. Elsadek[2], A. Abdel-Rahman[1], S. Hanif[3], R. K. Pokharel[3], [1]Egypt-Japan University of Science and Technology, New Borg-Alarab, Egypt, [2]Electronics Research Institute, Dokki, Egypt, [3]Kyushu University, Fukuoka, Japan

San Diego Trolley
Courtesy: Omni Hotel, San Diego

WEDNESDAY, 28 JANUARY 2015

WiSNET Session: WE3A	RWS Session: WE3B	RWS Session: WE3C	WiSNET Session: WE3D
Six-Port and Multi-Port Technology	**3D and Printed Technologies for RF**	**Late News II**	**Novel Sensors and Sensor Components**
Chair: Fadhel Ghannouci, University of Calgary Co-Chair: Alexander Koelpin, University of Erlangen Room: Gallery 1	Chair: Shamsur Mazumder Room: Grand Salon A	Chair: Sergio Pacheco, Freescale Co-Chair: Karen Gettings, MIT Lincoln Laboratory Room: Grand Salon B	Chair: Nils Pohl, Fraunhofer Institute for High Frequency Physics and Radar Techniques Co-Chair: Changzhi Li, Texas Tech University Room: Gallery 2

13:30

WE3A-1 Six-Port Technology for MIMO and Cognitive Radio Receiver Applications (Invited)

A. Hasan[1], M. Helaoui[1], N. Boulejfen[2,1], F. Ghannouchi[1], [1]University of Calgary, Calgary, Canada, [2]University of Hail, Hail, Saudi Arabia

WE3B-1 RCS Reduction of Ridged Waveguide Slot Antenna Array with Metamaterial Absorber

X. Cao, W. Li, J. Gao, Q. Yang, Air Force Engineering University, Xi'an, China

WE3C-1 High-Performance W-band LNA and SPDT Switch in 0.13 um SiGe HBT Technology

C. A. Ulusoy[1], R. Schmid[1], M. Kaynak[2], B. Tillack[2,3], J. D. Cressler[1], [1]Georgia Institute of Technology, Atlanta, United States, [2]IHP Microelectronics GmbH, Frankfurt (Oder), Germany, [3]Technische Universitaet Berlin, Berlin, Germany

WE3D-1 Millimeter-Wave Radar Systems On-Chip and in Package: Current Status and Future Challenges (Invited)

R. Feger, A. Stelzer, Johannes Kepler University Linz, Linz, Austria

13:50

WE3A-2 100 GHz Reflectometer for Sensitivity Analysis of MEMS Sensors Comprising an Intermediate Frequency Six-port Receiver

S. Linz, F. Oesterle, A. Talai, S. Lindner, S. Mann, F. Barbon, R. Weigel, A. Koelpin, Friedrich-Alexander University of Erlangen-Nuremberg, Erlangen, Germany

WE3B-2 2.4 GHz Inkjet-Printed RF Energy Harvester on Bulk Cardboard Substrate (Invited)

Z. Khonsari[1], T. Björninen[1], M. M. Tentzeris [2], L. Sydänheimo[1], L. Ukkonen[1], [1]Tampere University of Technology, Tampere, Finland, [2]Georgia Institute of Technology, Atlanta, United States

WE3C-2 A Broadband Rx Band Noise Reduction Circuit with CMOS Switch for Multi-Band Power Amplifier

Y. Kawamura, S. Shinjo, K. Iyomasa, M. Hirobe, K. Kato, Y. Takahashi, S. Yamabe, K. Horiguchi, M. Hieda, K. Yamanaka, Mitsubishi Electric Corporation, Kamakura, Japan

WE3D-2 A 7-µW 2.4-GHz Wake-Up Receiver with 80 dBm Sensitivity and High Co-Channel Interferer Tolerance

H. Milosiu, F. Oehler, M. Eppel, D. Fruehsorger, T. Thoenes, Fraunhofer IIS, Erlangen, Germany

14:10

WE3A-3 Forward V-band Vector Network Analyzer Based on a Modified Six-port Technique

K. Haddadi, T. Lasri, IEMN, Villeneuve d'Ascq, France

WE3B-3 Ultra-Wideband Microwave Components Fabricated Using Low-Cost Aerosol-jet Printing Technology

X. Lan[1], X. Lu[2], T. Blumenthal[3], V. Fratello[3], W. Chan[1], M. Truong[1], K. Kiyono[1], Y. Zhang[2], G. Gu[2], M. Tan[1], [1]Northrop Grumman, Redondo Beach, United States, [2]University of Massachusetts Lowell, Lowell, United States, [3]QI2, Kent, United States

WE3C-3 Transmit-Receive (T/R) Isolation Enhancement with an Indented Antenna Array

Q. Xu, S. Qin, Y. E. Wang, UCLA, Los Angeles, United States

WE3D-3 A Time to Digital Converter for use in Ultra Wide Band Radar Sensor Nodes

D. Genschow, IHP, Frankfurt (Oder), Germany

14:30

WE3A-4 A New Compact V-band Six-Port Down-Converter Receiver for High Data Rate Wireless Applications

C. Hannachi, S. Tatu, Institut National de la Recherche Scientifique-EMT, Montréal, Canada

WE3B-4 Planar Monopole Antennas on Substrates Fabricated Through an Additive Manufacturing Process

C. D. Saintsing[1], K. Yu[2], H. J. Qi[2], M. M. Tentzeris[1], [1]Georgia Institute of Technology(ECE), Atlanta, United States, [2]Georgia Institute of Technology(Mechanical), Atlanta, United States

WE3D-4 Generation of UWB pulses utilizing directly modulated tunable MEMS-VCSEL (Invited)

C. Gierl, Q. T. Le, C. Damm, F. Küppers, TU Darmstadt, Darmstadt, Germany

14:50

WE3A-5 ADC Depending Limitations for Six-Port Based Distance Measurement Systems

S. Lindner, F. Barbon, S. Linz, F. Lurz, S. Mann, R. Weigel, A. Koelpin, University of Erlangen-Nuremberg, Erlangen, Germany

WE3D-5 Diode Detector Design for 61 GHz Substrate Integrated Waveguide Six-Port Radar Systems

S. Mann, S. Erhardt, S. Lindner, F. Lurz, S. Linz, F. Barbon, R. Weigel, A. Koelpin, University of Erlangen-Nuremberg, Erlangen, Germany

WEDNESDAY, 28 JANUARY 2015

WiSNet Session: WE4A	RWW Session: WE4B	RWW Session: WE4D
Sensor Networks for Modern Applications	**Wireless System Modelling**	**Digital Signal Processing**
Chair: Christian Damm, University of Darmstadt Co-Chair: Dietmar Kissinger, IHP GmbH	Chair: Syed Islam, University of Tennessee at Knoxville	Chair: Jeremy Muldavin, MIT Lincoln Laboratory
Room: Gallery 1	Room: Grand Salon A	Room: Grand Salon B

15:40

WE4A-1 An Ultra-High Resolution Radar-System Operating at 300 GHz (Invited)

N. Pohl[1], S. Stanko[1], M. Caris[1], A. Tessmann[2], M. Schlechtweg[2], [1]Fraunhofer FHR, Wachtberg, Germany, [2]Fraunhofer IAF, Freiburg, Germany

WE4B-1 Phase Noise Cancellation Performance in Self-Heterodyning Transceivers for Wireless Backhaul Applications

S. Maier, X. Yu, H. Schlesinger, G. Luz, P. Jueschke, U. Seyfried, A. Pascht, Alcatel-Lucent Bell Labs Germany, Stuttgart, Germany

WE4D-1 High-Resolution Indoor Positioning System using SDR Modules

A. N. Gaber, S. Prcanovic, A. Omar, The University of Magdeburg, Magdeburg, Germany

16:00

WE4A-2 Millimeter-wave Radar Distance Measurements in Micro Machining

S. Ayhan[1], S. Thomas[2], N. Kong[3], S. Scherr[1], M. Pauli[1], T. Jaeschke[4], J. Wulfsberg[3], N. Pohl[2], T. Zwick[1], [1]Karlsruhe Institute of Technology, Karlsruhe, Germany, [2]Fraunhofer Institute for High Frequency Physics and Radar Techniques, Wachtberg, Germany, [3]Helmut Schmidt University - Hamburg, Hamburg, Germany, [4]Ruhr University Bochum, Bochum, Germany

WE4B-2 Enhancing Connectivity for Communication and Control in Unmanned Aerial Vehicle Networks

D. B. Rawat, R. Grodi, C. Bajracharya, Georgia Southern University, Statesboro, United States

WE4D-2 Outphasing Multi-Level RF-PWM Signals for Inter-Band Carrier Aggregation in Digital Transmitters

S. Chung[1,2], R. Ma[1], K. Teo[1], K. Parsons[1], [1]Mitsubishi Electric Research Laboratories, Cambridge, United States, [2]MIT, Cambridge, United States

16:20

WE4A-3 Structural Health Monitoring of Wind Turbines using Low-Cost Portable K-band Radar: an ab-initio Field Investigation (Invited)

T. Nikoubin[1], J. Muñoz-Ferreras[3], R. Gómez-García[3], D. Liang[2], C. Li[1], [1]Texas Tech University, Lubbock, United States, [2]Texas Tech University, Lubbock, United States, [3]University of Alcalá, Alcalá de Henares, Spain

WE4B-3 All-Analog Peak-to-Average Power Reduction using Constrained Clipping for OFDM Systems

M. Cho, J. S. Kenney, Georgia Institute of Technology, Atlanta, United States

WE4D-3 Frequency Quadrupling Transmitter Architecture with Digital Predistortion for High-Order Modulation Signal Transmission

Y. Liu, G. Liu, P. M. Asbeck, University of California at San Diego, La Jolla, United States

16:40

WE4A-4 Underwater Interferometric Radar Sensor for Distance and Vibration Measurement

M. Sporer[1], F. Lurz[1], E. Schluecker[2], R. Weigel[1], A. Koelpin[1], [1]University of Erlangen-Nuremberg (Inst. Elec. Eng.), Erlangen, Germany, [2]University of Erlangen-Nuremberg (Inst. Proc. Tech. and Mach.), Erlangen, Germany

WE4B-4 An Experimental Evaluation on EVM Performance for 4-CSK (Color Shift Keying) using Visible Light with Multiple Full-color LEDs

H. Shimamoto, Y. Kozawa, Y. Umeda, Tokyo University of Science, Noda, Japan

WE4D-4 Real Time Digital Signal Strength Tracking for RF Source Location

J. D. Popp[1], J. Lopez[2], [1]University of Washington, Seattle, United States, [2]NoiseFigure Research, Inc., Lubbock, United States

17:00

WE4A-5 Urban Highway Bridge Structure Health Assessments using Wireless Sensor Network

F. X. Li[1], A. Islam[2], A. S. Jaroo[2], H. Hamid[2], J. Jalali[1], M. Sammartino[1], [1]Youngstown State University, Youngstown, United States, [2]Youngstown State University, Youngstown, United States

WE4D-5 Digital Cancellation Technique to Mitigate Receiver Desensitization in Cellular Handsets Operating in Carrier Aggregation Mode With Multiple Uplinks and Multiple Downlinks

H. Gheidi[1], H. T. Dabag[2], Y. Liu[1], P. M. Asbeck[1], P. Gudem[2,1], [1]University of California San Diego, La Jolla, United States, [2]Qualcomm Inc, San Diego, United States

Industry Exhibits

Industry Exhibits	Exhibitor	Booth
Room: **Grand Salon CDE**	Sonnet Software, Inc (Diamond Sponsor)	3
	Keysight Technologies (Sponsor)	8
Monday, 26 January 2015 13:00 – 17:30	Virginia Diodes, Inc. (Sponsor)	13
	Berkeley Nucleonics	10
Tuesday, 27 January 2015 10:00 – 17:00	CST of America, Inc	11
	EMSCAN	15
	Focus Microwaves Inc	14
	Kyocera America, Inc	7
	Maury Microwave	6
	Microwave Product Digest	5
	MOSIS	2
	National Instruments	12
	Remcom, Inc	9
	RF Micropower	4
	West Bond Inc	1

Diamond Sponsor:

Sponsors:

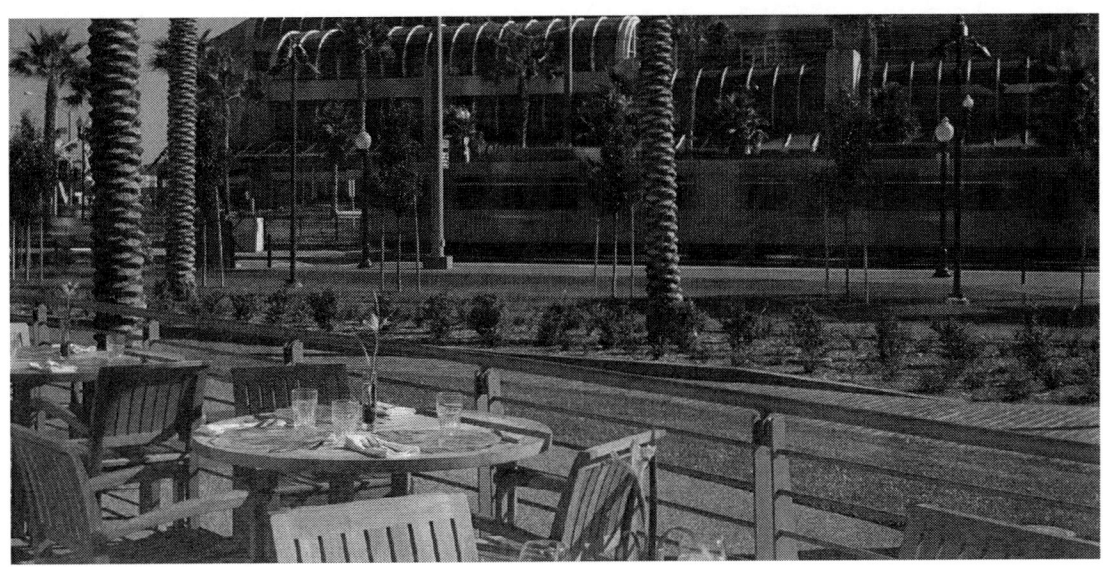

Omin Hotel, Mccormick Schmics Sea Food Restaurant
Courtesy: Omni Hotel, San Diego

Hotel Maps

4th Floor

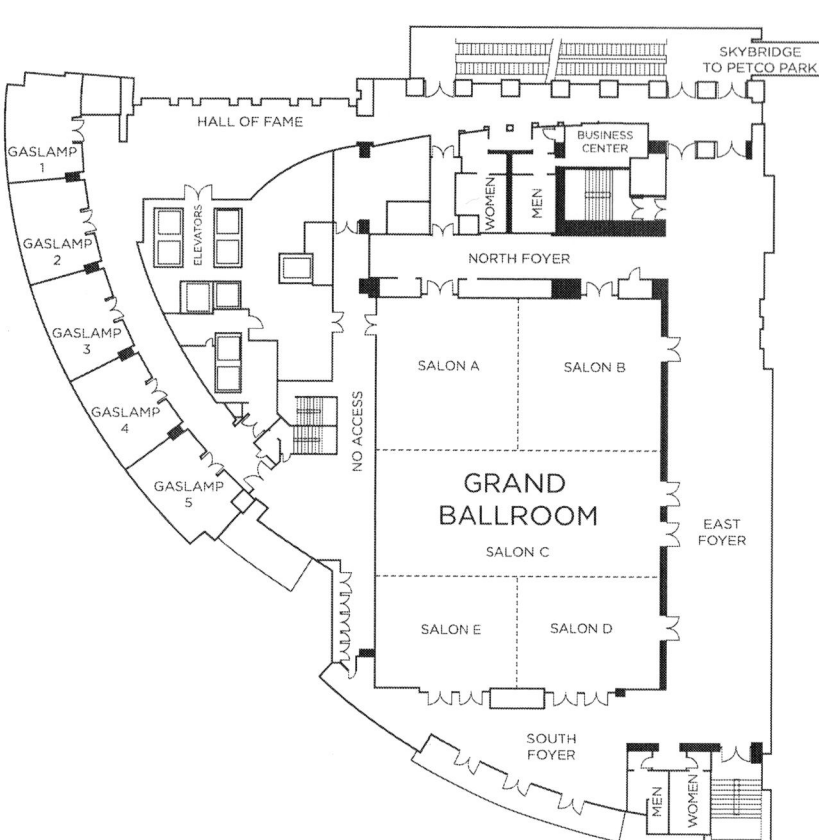

5th Floor

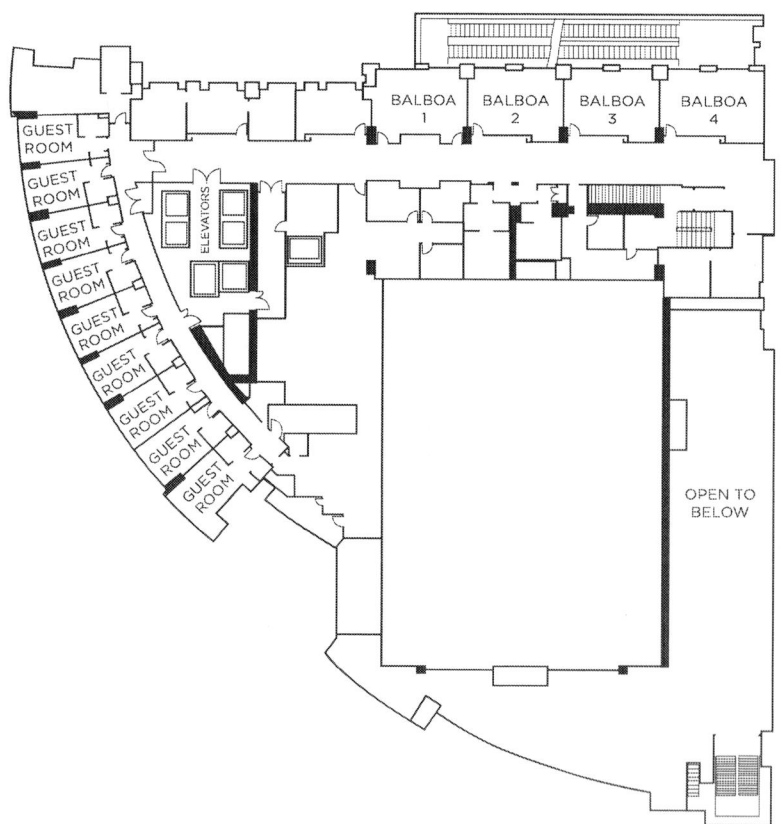

Gallery Meeting Space

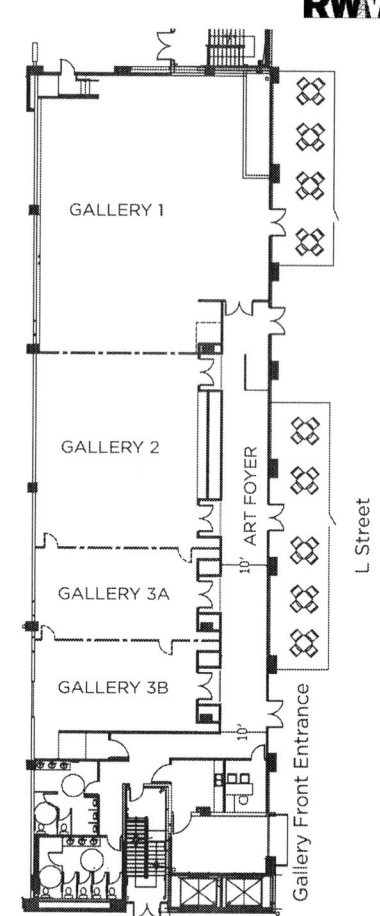

Driving Directions to the OMNI San Diego, CA

Address: 675 L Street, San Diego, CA 92101
Telephone: +1 (619) 231-6664 or
1-800-THE-OMNI (843-6664)

FROM SAN DIEGO INTERNATIONAL AIRPORT – 14 Min./4 Mi.
Follow the airport exit signs toward Harbor Drive/Downtown San Diego. Merge onto Harbor Drive going south along the San Diego Bay. Follow Harbor Drive as it turns to the left at Seaport Village. Turn left onto 5th Avenue. Make an immediate right onto L Street. Proceed 1 block on L Street. The hotel is located on the right on the corner of 6th Avenue and L Street.

DIRECTIONS FROM THE NORTH — VIA HIGHWAY 5 SOUTH
Take Highway 5 South to the 10th Avenue exit. Follow 10th Avenue to Market Street. Turn right onto Market Street and proceed to 6th Avenue. Turn left onto 6th Avenue and proceed to L Street. The hotel is on the left on the corner of 6th Avenue and L Street.

DIRECTIONS FROM THE NORTH — VIA HIGHWAY 163 SOUTH
Take Highway 163 until it turns into 10th Avenue. Follow 10th Avenue to Market Street. Turn right onto Market Street and proceed to 6th Avenue. Turn left onto 6th Avenue and proceed to L Street. The hotel is on the left on the corner of 6th Avenue and L Street.

DIRECTIONS FROM THE EAST — VIA HIGHWAY 8 WEST.
Take Highway 8 West to Highway 163 South toward downtown. Take Highway 163 until it turns into 10th Avenue. Follow 10th Avenue to Market Street. Turn right onto Market Street and proceed to 6th Avenue. Turn left onto 6th Avenue and proceed to L Street. The hotel is on the left on the corner of 6th Avenue and L Street.

DIRECTIONS FROM LOS ANGELES OR ORANGE COUNTY
Take Interstate 405 South until it merges with Interstate 5 South. Take Highway 5 South to the 10th Street exit. Follow 10th Street to Market Street. Turn right onto Market Street and proceed to 6th Avenue. Turn left onto 6th Avenue and proceed to L Street. The hotel is on the left on the corner of 6th Avenue and L Street.

RWW 2015 at a Glance

Activity	Location	Sunday (Jan. 25, 2015) Afternoon	Sunday Evening	Monday (Jan. 26, 2015) Morning	Monday Afternoon	Monday Evening	Tuesday (Jan. 27, 2015) Morning	Tuesday Afternoon	Tuesday Evening	Wednesday (Jan. 28, 2015) Morning	Wednesday Afternoon	Wednesday Evening
RWW Workshops	Gaslamp 1	13:30-17:30										
RWW Workshops	Gaslamp 2											
RWW Workshops	Gaslamp 3	13:30-17:30										
RWW Workshops	Gallery 1				13:30-17:30							
Panel	Grand Salon B					19:00-20:30						
Industry Forum	Gallery 3A			9:00-12:00								
RWW Plenary	Grand Salon AB						10:10-12:10					
RWS Sessions	Grand Salon A, Grand Salon B, Gallery 1, Gallery 2			8:00-9:20, 10:10-11:30	13:30-15:10, 15:40-17:20		8:00-9:00	13:30-14:50, 16:00-17:20		8:00-9:40, 10:10-11:50	13:30-14:50, 15:40-17:00	
PAWR Sessions	Gallery 1			8:00-9:40, 10:10-11:50	13:30-15:10, 15:40-17:20							
WiSNet Sessions	Gallery 1, Gallery 2									8:00-9:40, 10:10-11:50	13:30-15:10, 15:40-17:20	
BioWireleSS Sessions	Gallery 2						8:00-9:40	13:30-14:50, 16:00-17:20		8:00-9:40, 10:10-11:50		
SiRF Sessions	Grand Salon A, Grand Salon B, Gallery 2			8:00-9:40, 10:10-11:50	13:30-15:10, 15:40-17:20		8:00-9:40	13:30-14:50		8:00-9:40, 10:10-11:30		
Distinguished Lectures I & II	Gallery 2			8:00-9:20, 10:10-10:50								
Student Paper Contest	Grand Salon CDE				14:20-16:10							
Interactive Poster Sessions	Grand Salon CDE							14:55-16:55		12:55-14:30		
Exhibits	Grand Salon CDE				13:00-17:30		10:00-17:00					
RWW MicroApps	Grand Salon CDE							14:00-16:20				
Breakfast	Palm Terrace			7:00-8:00			7:00-8:00			7:00-8:00		
AM Coffee Break	Salon CDE			9:40-10:10			9:40-10:10			9:40-10:10		
PM Coffee Break	Salon CDE				15:10-15:40			15:10-15:40			15:10-15:40	
RWW Reception	Palm Terrace					18:00-20:00						
RWW Awards Banquet	Gallery 3B								18:30-21:00			

SIMMWIC Integration of Millimeter-Wave Antenna With Two Terminal Devices For Medical Applications

Erich Kasper and Wogong Zhang

University of Stuttgart, Institute for Semiconductor Engineering,

Pfaffenwaldring 47, D-70569 Stuttgart

kasper@iht.uni-stuttgart.

Abstract — Integration of planar antenna structures with oscillators and rectifiers, respectively, was performed an low loss, high resistivity silicon substrates. IMPATT diodes and special Schottky diodes operated in Mott mode were monolithically integrated in the planar waveguide circuit for mm-wave operation in the W-band (silicon monolithic mm-wave integrated circuit-SIMMWIC). Small chip sizes below 30mm^2 could be obtained to meet requirement for medical endoscopes.

Index Terms — mm-wave, planar antenna, SIMMWIC, IMPATT diode, Schottky diode, Mott operation, endoscope, medical diagnosis

I. INTRODUCTION

A broad electromagnetic wavelength spectrum is used in medicine and biology [1, 2] to investigate the human body and tissues mainly for diagnosis and partly also to affect biological processes. These investigations are limited in many cases to the outer shell because of size limitations for inside diagnosis. In inside diagnosis we see two directions of use: (i) Endoscope with sensors in their head. The full sensor size is limited to a few millimeters. The sensor information is delivered to the outside monitor and preferably processed in real time. (ii) Autonomous capsules which either store the sensor information for later processing in a memory or are connected to the monitor with radio link. The usual sensor in the endoscope head is an optical camera but additional microwave sensors will enhance the diagnostic abilities [3] considerably. Microwaves up to the terahertz regime penetrate about around a mm into different tissues giving subsurface information for tissue identification and disease detection. Range information from the reflected beam is highly useful for the safe manipulation of the endoscope.

Endoscope heads have a maximum diameter of one centimeter and the tendency is to shrink that size. The full microwave sensor has to fit on that front head. An absolute priority in medical applications for in body diagnosis or treatment is the size limitation for the complete microwave front end to a few square millimeters. Frequencies in the mm-wave range are preferred because of smaller wavelength. That means a full integration of receiver and transmitter electronics with mm-wave antenna is necessary.

The recent progress [4] in silicon based technologies for mm-wave generation and detection allow to propose integrated solutions for medical purposes. We propose and investigate in this paper an integration technique based on low loss, high resistivity silicon substrates. This SIMMWIC technique [5] is targeted for small size and low power consumption, and it will be scalable down to sub-millimeter wavelengths.

II. SILICON MONOLITHIC MILLIMETER-WAVE INTEGRATED CIRCUIT

Silicon substrate material with a high specific resistivity up to 10,000 cm is available from commercial substrate suppliers. Planar mm-wave circuits on these substrates show negligible substrate loss if metal oxide semiconductor (MOS) inversion or accumulation carrier sheets below the signal lines are avoided [6] by proper biasing or technical measures like surface treatments. RF systems based on active integrated antennas [7] are demonstrated to be realizable. Transistors (silicon germanium hetero-bipolartransistor HBT, RF-MOS) and specific diodes (Schottky barrier diodes, impact avalanche transit time IMPATT diodes) are available for operation of microwave oscillators, amplifiers and detectors. We realized diode driven circuits for the benefit of simple and small circuits. As these diodes are based on Schottky and p/n junctions they are compatible with CMOS technology [8]

III. MOTT DIODE RECTIFIERS

Usual Schottky diodes exhibit an increasing capacitance $C_j(V)$ with forward voltage which is the chosen operation regime for rectification. Schottky diodes with low doped epitaxial layers (Mott diodes) are preferred for rectification purposes because the junction capacitance C_j stays constant (C_{j0}) up to a threshold voltage which is equal the built-in voltage V_{bi} of the junction in ideal Mott diodes but smaller in real Mott diodes with finite background doping.

978-1-4799-8198-4/15 $31.00 © 2015 IEEE

We used vertical Mott diode with nominal 250nm undoped (n-background doping below $10^{15}/cm^3$) Si layers on p$^+$ buried layer for backside contacts. The front side Schottky contact, the Ohmic buried layer contact and the planar waveguide structures were fabricated with aluminum as metallization. .

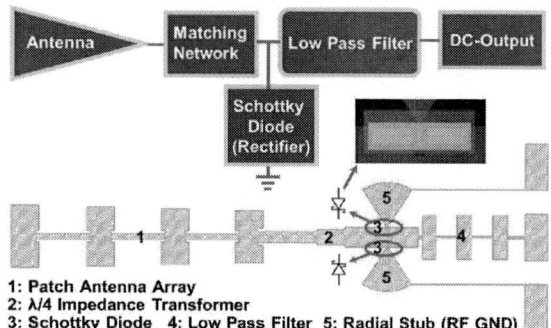

1: Patch Antenna Array
2: λ/4 Impedance Transformer
3: Schottky Diode 4: Low Pass Filter 5: Radial Stub (RF GND)

Fig.1 SIMMWIC Schottky diode rectifier

The structuring of the diodes, the waveguides and antenna was performed with a low thermal budget (< 400°C) process. As integrated receiver circuit we have chosen a rectifying antenna (rectenna) (Fig.1) with a projected operation frequency of 85GHz. The rectification is measured as voltage change ΔV vs frequency for a given forward bias current I_{bias} (Fig.2).

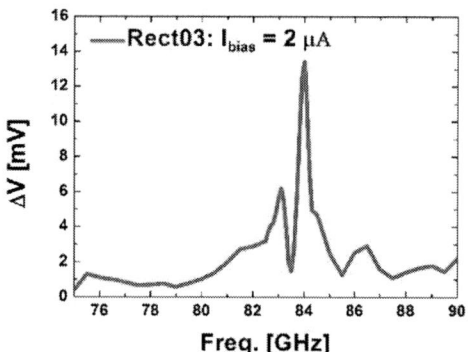

Fig.2 Spectral response of the rectifier with a patch antenna

The rectifying properties of Schottky diodes in general are defined by their exponential current-voltage characteristics. The small signal equivalent circuit is composed of the inner diode with parallel conductance G_p and junction capacitance C_j and the series resistance R_s. The high frequency losses stem from the voltage drop across the series resistance R_s and the non-rectifying current path along the junction capacitanc C_j . The incoming AC wave from the antenna has a power P proportional to the product $i \cdot u$ with voltage and current amplitudes, u and i respectively. The power Pp in the rectifying branch (represented in the equivalent circuit by

the characteristics dependent value G_p) is smaller and given proportional to the product i_p with u_j.

$$\left|\frac{P_p}{P}\right|^2 = \frac{|u_j i_p|^2}{|ui|^2} = \left[1 + (\frac{\omega}{\omega_p})^2\right]^{-1} \left[(1+\frac{\omega_p}{\omega_s})^2 + (\frac{\omega}{\omega_s})^2\right]^{-1} \quad (1)$$

using the abbreviations ω_p and ω_s for parallel cut off frequency and serial cut off frequency, respectively.

$$\omega_p = \frac{G_p}{C_j}; \omega_s = \frac{1}{R_s C_j} \quad (2)$$

Equ. (1) shows that only a part of the incoming power is directed toward the rectifying branch of the diode.

The frequency dependence of the rectifying property is much more complicated than only described by a single cut off frequency as often is assumed.

In order to investigate this frequency behavior we consider a fixed operation frequency $f = f_0$ (85GHz in our case) and the characteristics of an ideal Mott diode.

$$G_p = \frac{I}{V_T} \quad (3)$$

Under this idealizing conditions we find a constant serial cut off ω_s and a forward current dependent parallel cut off

$$\omega_p = \frac{I}{V_T C_j}; \quad (4)$$

The parallel cut off frequency ω_p occurs in Equ. (1), both in the nominator and denominator terms giving an optimum power efficiency at $\omega_{p,max}$. For $\omega_{p,max}$ we obtain the following implicit equation (equ.1)

$$\omega_{p,max}^4 + \omega_s \cdot \omega_{p,max}^3 - \omega_p \cdot \omega_s \cdot \omega_0^2 - \omega_s^2 \cdot \omega_0^2 - \omega_0^4 = 0 \quad (5)$$

The calculated optimum frequency ω_p and the optimum power efficiency from equ.1 is shown in Fig.3.

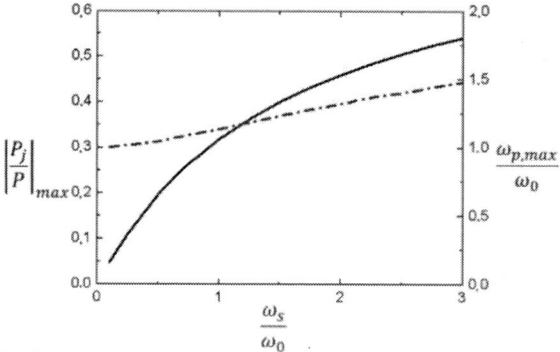

Fig.3 Optimum parallel cut off frequency ω_p (broken line) and corresponding optimum power efficiency $|P_j/P|_{max}$ vs serial cut off frequency ω_s for an ideal Mott diode.

The experimental dependence of the voltage change ΔV as function of the current shows the predicted maximum behavior (equ.1) but at lower current levels as expected from equ.5 (Fig.4). This is probably caused by the

978-1-4799-8198-4/15 $31.00 © 2015 IEEE

deviation from ideality of real Mott diode capacitance below the built-in voltage V_{bi}.

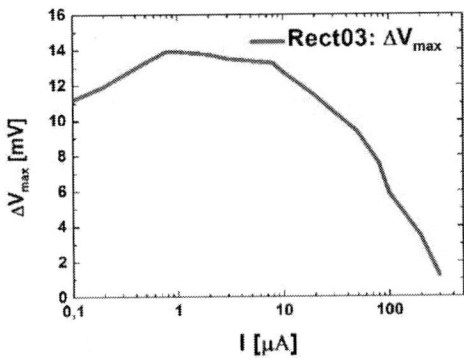

Fig.4 Rectifier voltage ΔV vs current I

IV. IMPATT TRANSMITTER

The layout of the IMPATT transmitter is shown in Fig.5.

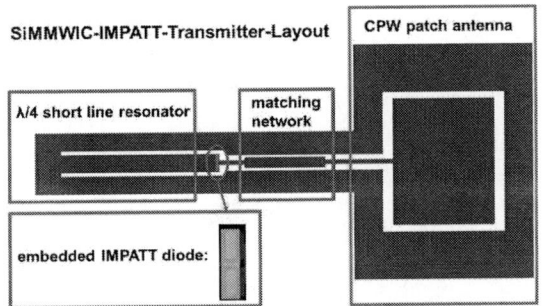

Fig.5 Layout of an IMPATT transmitter

It consists of the coplanar waveguide (CPW) antenna, a resonator, the monolithically embedded IMPATT diode [9] and a matching network. S-Parameter response and beam characteristics of the CPW antenna on backside metallized Si substrate are simulated with ADS and CST Microwave Studio (Fig.6).

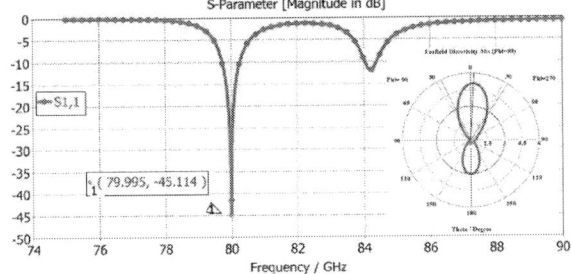

Fig.6 Simulation of the integrated CPW antenna

The resonance frequency (avalanche frequency) of the IMPATT diode has to be shifted near the operation frequency by appropriate bias current. S-Parameter

measurements of the IMPATT diode reveal the resonance with zero imaginary impedance part Im(Z) and high negative real impedance part Re(Z). At the chosen bias current of 40mA the avalanche frequency is shifted to 95GHz (Fig.7).

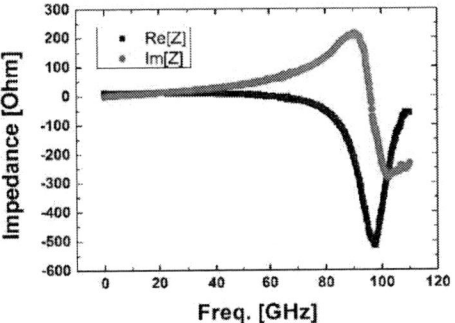

Fig.7 Impedance of the IMPATT diode (bias 40 mA)

V. CONCLUSIONS

Integrated mm-wave receivers and transmitter (W-band) with small area consumption (IMPATT transmitter: $5mm^2$, Mott diode Rectenna: $< 30\ mm^2$) were designed and realized in SIMMWIC technology. Future work will concentrate on a single chip transceiver with an even more reduced area by a compact CPW design.

ACKNOWLEDGEMENT

We acknowledge discussions and technical assistance from M. Oehme, K. Kostecki, K. Matthies, S. Rohmer, W. Kasper, M. Model and K. Ye.

REFERENCES

[1] C.Gu and C. Li, From tumor targeting to speech monitoring, IEEE Microwave Magazine 15, no.4, 66 (2014)

[2] N.K. Nikolova, Microwave imaging for breast cancer, IEEE Microwave Magazine 12 ,no.6, 78 (2011)

[3] P.H. Siegel, "Terahertz technology in biology and medicine", *IEEE Trans. MTT* 52, 2438(2004).

[4] E. Kasper, D. Kissinger, P. Russer, and R. Weigel, "High speeds in a single chip", *IEEE Microwave Magazine* 10, 28(2009).

[5] P. Russer, "Si and SiGe mm-wave integrated circuits", *IEEE Trans. MTT* 46, 590(1998).

[6] C. Schoellhorn, W. Zhao, M. Morschbach and E. Kasper, "Attenuation mechanisms of Al mm-wave coplanar waveguides on Si", *IEEE Trans. ED* 50, 740(2003).

[7] E. Biebl, "RF systems based on active integrated antennas", *Int. J. Electron. Commun.* 57, 173(2003).

[8] T. Al-Atar and T.H. Lee, "Monolithic integrated mm-wave IMPATT transmitter in standard CMOS technology", *IEEE Trans. MTT* 53, 3557(2005).

[9] M. Oehme, E. Kasper, "IMPATT diodes", *in Silicon Heterostructure Handbook (Ed. D. Cressler)*, CRC Press, 661(2006).

A SiGe Differential 50 ps Gaussian Pulse Generator for Sub-Sampling TDR Measurements

Gregor Hasenaecker[1], Hans-Martin Rein[1], Klaus Aufinger[2], Nils Pohl[3], and Thomas Musch[1]

[1]Ruhr-Universität Bochum, 44780 Bochum, Germany
[2]Infineon Technologies, 85579 Neubiberg, Germany
[3]Fraunhofer FHR, 53343 Wachtberg, Germany

Abstract— **A high-speed pulse generator with two complementary outputs has been realized in a SiGe bipolar technology. Both ultra-short output pulses (with different polarity) are well approximated by a Gaussian function and show a width t_p of about 50 ps and an amplitude of 3.4 V, resulting in an amplitude at the differential output of 6.8 V. Such short pulses are essential in sub-sampling TDR systems as t_p limits the bandwidth and thus the resolution of the measurement applications. The monolithically integrated pulse generator presented here shows substantial performance improvement compared to already existing solutions, mostly based on step recovery diodes.**

I. INTRODUCTION

Time-domain reflectometry (TDR) systems emit short electromagnetic pulses to a measurement object and evaluate the reflections to obtain information about distance or material characteristics. Established applications are industrial liquid-level measurements [1] and process diagnostics in plasma. For instance, with the concept of a multipole resonance probe (MRP) described in [2], relevant plasma parameters can be extracted if the plasma is excited in a frequency range from DC to 5 GHz.

One crucial component in these TDR systems is a pulse generator, that is designed to generate signals with high bandwidth and output power to cover the demanded frequency and dynamic range, respectively.

Short electromagnetic Gaussian pulses (see Fig. 1)

$$v_g(t) = V_p \exp\left(-\frac{t^2\,4\ln 2}{t_p^2}\right) \quad (1)$$

with amplitude V_p and pulsewidth t_p (full width at half the maximum, FWHM) are particularly qualified for these applications. Due to the equal form of the Gaussian function in time- and frequency domain, there is a simple relationship for the pulsewidth t_p and the 3 dB bandwidth Δf of Gaussian pulses:

$$\Delta f = \frac{0.31}{t_p}. \quad (2)$$

This means that for the above mentioned applications with $\Delta f = 5$ GHz a pulsewidth $t_p \leq 62$ ps is required.

Due to the very small pulsewidth a direct sampling of the reflections is not feasible. Therefore, a sub-sampling

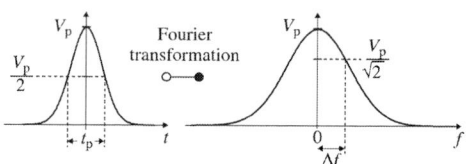

Fig. 1. Definitions of pulsewidth t_p and 3 dB bandwidth Δf for a Gaussian pulse in time- and frequency domain, respectively.

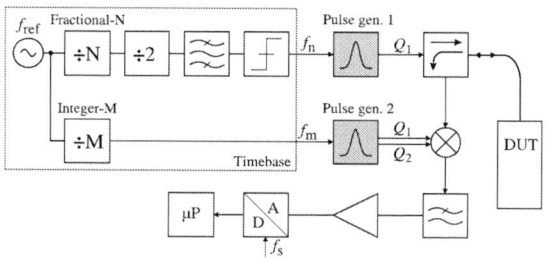

Fig. 2. Block diagram of a TDR measurement system for evaluation of reflections with respect to distance or material characteristics [3]. From pulse generator 1 only a single output is used, while the mixer is driven by the differential output of pulse generator 2.

mixer is usually applied. A balanced mixer is preferred to obtain a good noise performance. This requires a differential pulse generator as a local oscillator. A block diagram of the utilized sub-sampling TDR concept is shown in Fig. 2. For accurate sub-sampling, the timebase has to achieve a high frequency resolution of only a few hundred Hertz (between f_m and f_n) as well as an extremely low jitter below 1 ps. The highly stable timebase is based on a simple, but very-low-noise concept, which makes use of a fractional-N frequency divider. Its realization and a prototype circuit of the pulse generator have been published in [3]. Compared to [3], the generator circuit presented in this paper is based on an advanced design and features more than twice the output amplitude as well as shorter and cleaner pulses.

Many conventional realizations of comb generators use step-recovery diodes (SRD), achieving pulsewidths down to 100 ps. For an overview s. Table 2 in [4]. Besides a poor availability of ultra-fast SRDs, these passive devices require high effort for isolation of input and output and also high driving power. A pulse generator built of discrete bipolar transistors overcomes these drawbacks, but at a minimum pulsewidth $t_p \approx 90\,\text{ps}$ and high effort for complementary pulse polarity [1]. This is why we preferred a monolithically integrated solution based on high-speed bipolar transistors in a differential current switch.

II. CIRCUIT DESIGN AND SIMULATION

The circuit is designed on the base of Infineon's $0.35\,\mu\text{m}$ SiGe bipolar production technology B7HF200 [5]. The transistors with cutoff frequencies of $f_T = 170\,\text{GHz}$ and $f_{max} = 250\,\text{GHz}$ feature fast switching speed and a sufficient current density of $j_C = 5\,\text{mA}/\mu\text{m}^2$ at maximum f_T, but quite low breakdown voltages (here guaranteed: $BV_{CB0} \approx 5.8\,\text{V}$ and $BV_{CE0} \approx 1.4\,\text{V}$), typical for high speed technologies.

We use the cascoded differential topology shown in Fig. 3. The internal signal v_i is derived from an external source and sharpened on chip so that its falling edge causes a fast transfer of the source current I_0 ($\approx 190\,\text{mA}$) from transistor T_1 to T_2. An LC-network ($L_{1,2} - C_{1,2} - L_{Q1,2}$) driven by the common-base stages (T_{K1}, T_{K2}) results in a differentiation of the current slope and affects pulse shaping. A delayed pattern of the input signal is used to turn on and off the source current I_0 right before and after the generation of the output pulses. This measure ensures that v_i's rising edge does not cause an undesired output pulse. Moreover, it reduces power consumption.

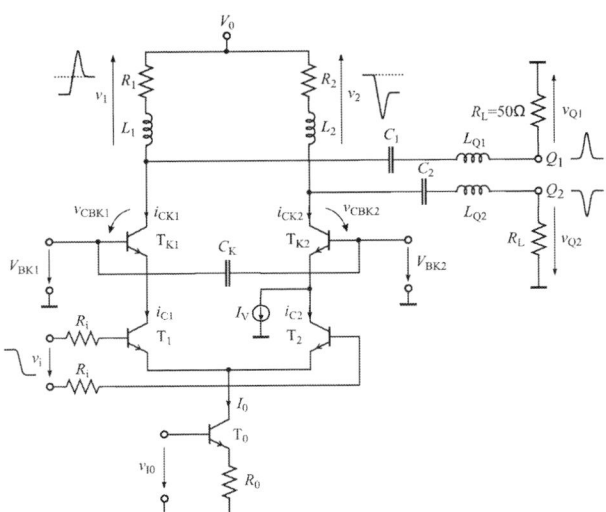

Fig. 3. Simplified circuit schematic of the pulse generator. Both v_i and v_{i0} are derived from a single external signal source using adequate on-chip circuits for shaping and delaying these signals. Furthermore, all bias voltages and current sources are located on-chip. I_V shortens the turn-on of T_{K2}.

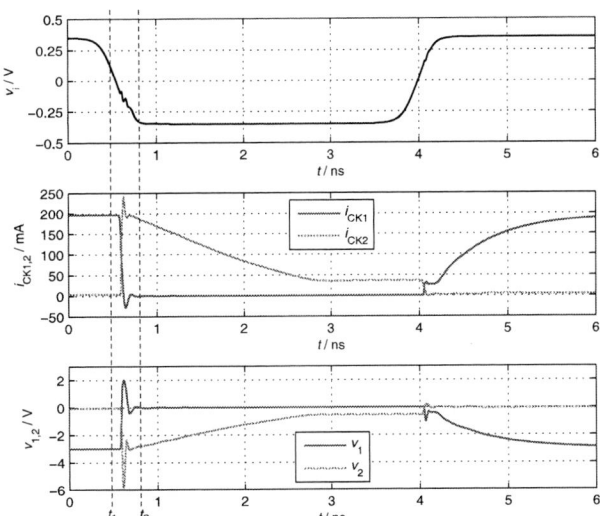

Fig. 4. Simulation results to illustrate the operating principle. The internal pulses v_1, v_2 (see Fig. 3) are generated during the falling edge of v_i and AC coupled to the output.

The simulation results in Fig. 4 help to understand the operation of the circuit in Fig. 3.

A main challenge of the design is to aim at high voltage amplitude at both outputs, despite the low BV_{CB0} of the transistors. This means, that the safe transistor operating range must be exhausted as far as possible, which is limited here mainly by the avalanche induced transistor breakdown. Even in the favorable common-base configuration, the corresponding CB-breakdown voltage can become much lower than BV_{CB0} since it decreases with increasing I_C, as a result of current pinch in. For details and accurate modeling see [6].

Fig. 3 shows, that the common-base stages T_{K1} and T_{K2} need different biasing V_{BK1} and $V_{BK2}(< V_{BK1})$ due to the different polarities of both output pulses. The resistors R_1,

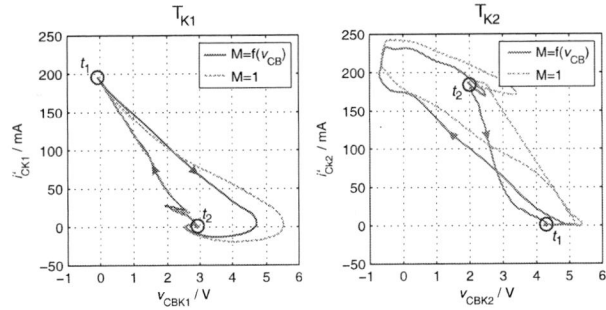

Fig. 5. Simulated dynamic output characteristics of the transistors T_{K1} and T_{K2} (inner collector current i'_C vs. the collector-base voltage v_{CB}). The output pulses are generated, when the circuit switches from the quasi-static operating point t_1 to t_2 according to Fig. 4.

R_2 are required to reduce the CB-voltage of a conducting transistor due to the current dependent breakdown behavior. However, they tend to degrade pulse performance and must therefore be carefully designed, together with the pulse shaping components and bias voltages in Fig. 3.

The simulation results in Fig. 5 show the output characteristics of the common-base stage transistors T_{K1}, T_{K2} during switching. The curves range from the high-current limit (left side) to near the guaranteed BV_{CB0} (right side), thus exhausting the usable transistor range to get maximum voltage swing. For this design, the multiplication factor was assumed to depend on v_{CB} only (worst case). For comparison, also the plots without multiplication ($M = 1$) are inserted.

The positive pulse v_{Q1} is slightly worse compared to v_{Q2}, mainly due to the severe influence of M near the end of the turn-off transient of T_{K1}. This effect is partly compensated by modifying the pulse shaping network L_1, L_{Q1}, C_1.

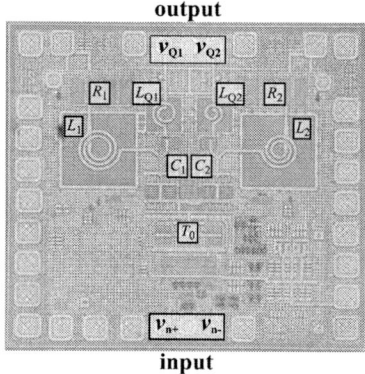

Fig. 6. Photo of the pulse generator chip ($\approx$ 1x1 mm^2) with special regard of the pulse shaping components in Fig. 3.

III. Experimental Results

The chip in Fig. 6 was fabricated in Infineon's SiGe bipolar production technology, shortly characterized in Chapter II, and mounted in a simple low-cost socket (teflon substrate on a brass block), using conventional wire bonding. The output signals are measured with a 50 GHz sampling scope (HP 54120B). The results, plotted in Fig. 7(a), are averaged over 1024 cycles with a pulse repetition frequency of 10 MHz. The signals at both outputs have an amplitude $V_p \approx \pm 3.4$ V and a pulsewidth $t_p \approx 50$ ps, resulting in an amplitude at the differential output ($v_{Q1} - v_{Q2}$) of 6.8 V at the same t_p. The sum and difference of both output signals is calculated in Fig. 7(b) and transformed to the frequency domain in Fig. 7(c) and (d). This points out the good suppression of the common mode output signal $v_{Q1} + v_{Q2}$ of more than 20 dB within the 3 dB bandwidth Δf. In the time- as well as the frequency-domain good approximation by the Gaussian function is observed. Additional measurements of the output signal

v_{Q1} and v_{Q2} by a 50 GHz spectrum analyzer (not shown) confirm the results from Fig. 7(c) and (d).

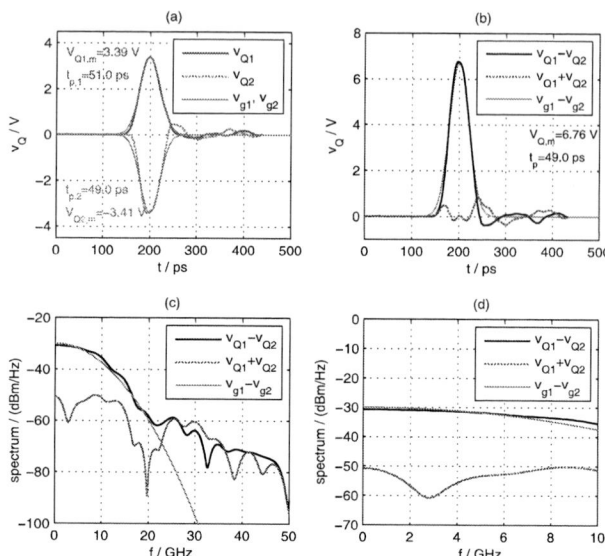

Fig. 7. Measured output pulses v_{Q1} and v_{Q2} (a) and derived differential and common-mode output pulse (b). The calculated spectra of $v_{Q1} \pm v_{Q2}$ in (c) and (d) point out $\approx$ 20 dB suppression of the common-mode output signal within the 3 dB bandwidth. All measurements fit well to ideal Gaussian pulses v_g plotted in the same figure.

IV. Conclusion

The realized pulse generator proved to be well suited for the intended application in an advanced TDR system for process diagnostics. The features achieved significantly exceed those of other published solutions. The amplitudes of both complementary outputs are ± 3.4 V resulting in 6.8 V at the differential output. The approximately Gaussian pulses show a width of 50 ps for a mounted chip.

References

[1] M. Gerding, T. Musch, and B. Schiek, "A novel approach for a high-precision multitarget-level measurement system based on time-domain reflectometry," *IEEE Transactions on Microwave Theory and Techniques*, vol. 54, no. 6, pp. 2768–2773, June 2006.

[2] M. Lapke et al., "The multipole resonance probe: Characterization of a prototype," *Plasma Sources Science and Technology*, vol. 20, no. 4, 2011.

[3] G. Hasenaecker, R. Storch, N. Pohl, and T. Musch, "A DC to 5 GHz TDR pulse generating unit with a highly stable timebase for use in a plasma diagnostic measurement system," *IEEE Int. Conference on Wireless Information Technology and Systems (ICWITS)*, Nov 2012.

[4] S. Oh and D. Wentzloff, "A step recovery diode based UWB transmitter for low-cost impulse generation," *IEEE Int. Conference on Ultra-Wideband (ICUWB)*, 2011.

[5] R. Vytla et al., "Simultaneous integration of SiGe high speed transistors and high voltage transistors," *Bipolar/BiCMOS Circuits and Technology Meeting (BCTM)*, Oct. 2006.

[6] M. Rickelt and H.-M. Rein, "A novel transistor model for simulating avalanche-breakdown effects in Si bipolar circuits," *IEEE Journal of Solid-State Circuits*, vol. 37, no. 9, pp. 1184–1197, Sep 2002.

A 6.5mW, Wide Band Dual-Path LC VCO Design With Mode Switching Technique in 130nm CMOS

Jun Li*, Ni Xu, Yuanfeng Sun, Woogeun Rhee and Zhihua Wang

Institute of Microelectronics, Tsinghua University, Beijing, China

*Now with University of California, San Diego, La Jolla, CA 92093 USA

Abstract—This paper demonstrates a dual path LC voltage controlled oscillator (VCO) that achieves ~65% tuning range with optimized inductor switching design. Combination of discrete and continuous tuning method reduces the gain ratio between coarse tuning path and fine tuning path which alleviates the noise and coupling issues due to high coarse tuning gain. Optimized inductor switching topology enables wideband continuous tuning from 2.74-5.37GHz. The VCO consumes 6.5 mW power from a 1.2 V supply and exhibits -123.8dBc/Hz at 3MHz offset from a 4.605 GHz carrier showing FOM_T of -198.3dBc/Hz. The circuit is fabricated in 130nm CMOS process and occupies an area of 0.6mm².

Index Terms — Dual-Path VCO, LC-VCO, Software Defined Radio (SDR), Cognitive Radio (CR), phase lock loops (PLL), voltage controlled oscillators (VCOs)

I. INTRODUCTION

The increasing demand for wide band wireless service is continuously generating new standards which targets for efficient spectrum usage. For instance, advanced mobile phones has enabled long-term evolution-advanced (LTE-A) system that adopts carrier aggregation technique to expand bandwidth. It can aggregates up to five 20-MHz channels from inter-bands or intra-bands [1]. IEEE 802.22 wireless regional area network (WRAN) uses cognitive-radio (CR) technique to exploit white spaces in the TV spectrum [2]. In addition, cognitive radio (CR) [3-4] technique needs to enable the spectrum sensing feature with programmability over the whole wireless bands for the long term development. According to specifications, smartphones, TV and WLAN require an operating range from 50MHz to 5GHz and cognitive radio (CR) technique might push it to even higher frequency in the near future. To achieve this wide operating range, one of the major challenges is the frequency synthesizer which provide local oscillators (LOs) for up-and down-conversion of modulated signals in RF transceivers. Furthermore, the main bottleneck of this synthesizer design is the wide range voltage controlled oscillator (VCO).

This work investigates the low power implementation of wide range voltage controlled oscillator (VCO). A dual-path, dual modes, LC VCO is proposed. By using inductor switching topology, the tuning range is up to ~65% without significant tank Q degradation. To overcome the inherent problem of having a very high gain ratio between the coarse-tuning path and the fine-tuning path, a partitioned coarse-tuning control is considered for low-noise over the wide frequency range.

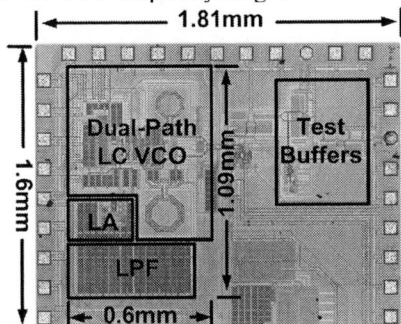

Fig. 1. Diephoto of proposed voltage control oscillator (VCO).

II. CIRCUIT ARCHITECTURE

Fig.1 shows the chip microphotograph of the proposed VCO. Here we focus on the oscillator design. Area efficient synthesizer design has been discussed in [5]. A phase-frequency detector (PFD) and a charge pump (CP) with a passive lowpass filter are used to generate the control voltage of the single input dual path VCO, where the control voltage range can be controlled by the dc gain of the linear amplifier (LA) in the coarse tuning path [6]. Since the tuning voltage of the fine-tuning varactor is near the middle of the tuning curve with a very small control voltage range as shown in Fig.2 The dual path PLL can achieve good CP linearity as well as small VCO gain variation.

Previous work [7] has solved the noise coupling issue in the high gain coarse-tuning path which affects the VCO noise performance. The coarse-tuning gain is reduced from 1GHz/V to 125MHz/V with 14dB linear amplifier (LA) gain. In this work, besides 5b capacitor bank control signal, an additional 1b inductor switching signal is proposed to provide dual-band wide range continuous tuning. The LA is designed with reconfigurable gain from 6dB to 18dB. K_{vco} is optimized to ~315MHz/V and ~70MHz/V for low phase noise in high band and low band. 64 discrete banks achieves continuous frequency tuning from 2.74GHz to 5.37GHz. Therefore, 50MHz to 5GHz wide band LO can be taken from PLL VCO buffer or divider buffers. Fig.2 illustrates the dual-band operation with inductor switching.

In high band mode, the inductor L_2 is enabled and the series capacitors are shorted by the switches M_6 and M_7. Then, the effective tank inductor becomes $L_1//L_2$ that shifts the resonant frequency to higher side. In low band mode, the inductor L_2 is disabled by complementary switch pair M_5 and M_{10} and the effective tank inductor becomes L_1.

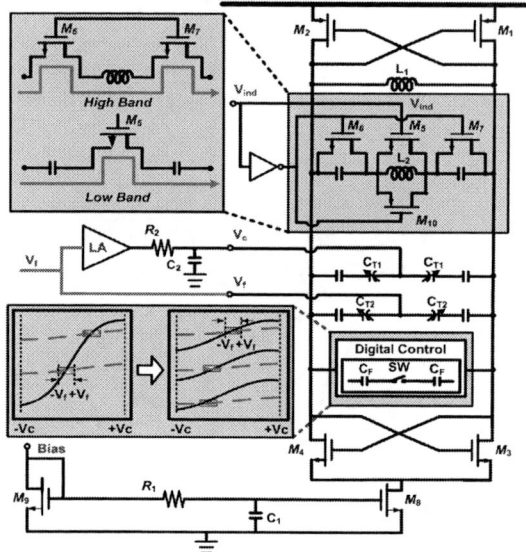

Fig. 2. Schematic of proposed dual-path mode switching VCO.

This work also investigates the design trade-off between tank Q and tuning range. Fig.3 shows the tank Q model in two modes operation. Where r_{low} and r_{high} are the equivalent low band and high band tank resistances. r_{L1} and r_{L2} are the parasitic inductor resistances. $r_{on,sw1}$ and $r_{on,sw2}$ are the switch resistance when they are turned on.

$$r_{low} \cong r_{L_1} + r_{on,sw_1} \qquad (1)$$

$$r_{high} \cong \frac{r_{L_1} + (r_{L_2} + r_{on,sw_2})(L_1 / L_2)^2}{(1 + L_1 / L_2)^2} r_{L_1} + r_{on,sw_1} \qquad (2)$$

Describing the capacitor load by C_{load}, the tank Q in tow modes can be expressed as,

$$C_{load} = C_{min} + \Delta C \qquad (3)$$

$$Q_{low} \cong \frac{1}{r_{low}} \sqrt{\frac{L_1}{C_{load} + C_1 / 2}} \qquad (4)$$

$$Q_{high} \cong \frac{1}{r_{high}} \sqrt{\frac{L_1 L_2}{C_{load}(L_1 + L_2)}} \qquad (5)$$

Where Q_{high} and Q_{low} are the quality values of the VCO tank in high band and low band. Since r_{high} is usually larger than r_{low}, in order to keep the quality value similar,

additional capacitor C_1 is inserted to balance the Q difference between two bands.

In order to achieve the continuous tuning, the overlap between high band minimum frequency $\omega_{min,high}$ and low band maximum frequency $\omega_{max,low}$ need to be insured. By using minimum capacitor load C_{min} and maximum capacitor load C_{max}, these two frequencies can be calculated. Where $\omega_{0,low}$ and $\omega_{0,high}$ are the low band and high band oscillation frequency.

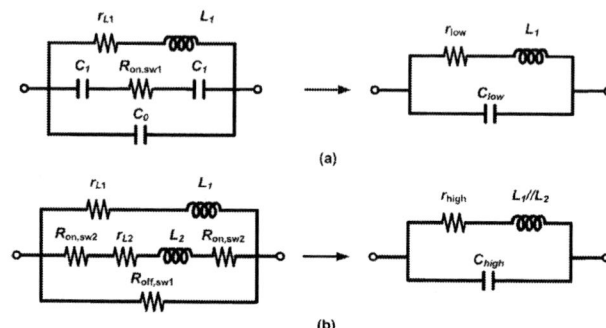

Fig. 3. (a) Low band and (b) High band small signal model

$$\omega_{0,low} \cong \sqrt{\frac{1}{L_1(C_{load} + C_1 / 2)}} \qquad (6)$$

$$\omega_{0,high} \cong \sqrt{\frac{1}{(L_1 // L_2)C_{load}}} \qquad (7)$$

Setting $\omega_{min,high} \leq \omega_{max,low}$ for continuous tuning, the inductor ratio L_1/L_2 needs to meet

$$\frac{L_1}{L_2} \geq \frac{C_{min} + C_1 / 2}{\Delta C - C_1 / 2} \qquad (8)$$

Since $r_{high} > r_{low}$, the ratio Q_{high} /Q_{low} can be expressed as,

$$\frac{Q_{high}}{Q_{low}} \cong \frac{r_{low}}{r_{high}} \sqrt{\frac{1 + \dfrac{C_1}{2C_{load}}}{1 + \dfrac{L_1}{L_2}}} \qquad (9)$$

Setting $Q_{high} \geq Q_{low}$ to keep the phase noise performance at high frequency, the design boundary is achieved.

$$\frac{C_1}{2C_{load,max}} \geq \frac{L_1}{L_2} \geq \frac{C_{min} + C_1 / 2}{\Delta C - C_1 / 2} \qquad (10)$$

III. MEASUREMENTS

To measure the phase noise of oscillator, a couple of output buffers are placed after VCO and dividers. The measured tuning range is shown in Fig.5 which covers 2.74GHz to 5.37GHz. With the help of divider chain in

TABLE I
VCO PERFORMANCE SUMMARY AND COMPARISON

Reference	This work	[3]	[4]	[6]	[7]	[8]	[9]	[10]
Technology (CMOS)	130nm	65nm	180nm	90nm	65nm	130nm	65nm	40nm
Frequency Tuning Rang (%)	~65	~136.8	~46.7	24.9	19.1	45	75	88.7
Phase Noise (dBc/Hz @ 1MHz)	-116.9	<-110	-120	-108	-124	-117.6	-119	-121.78
Power (mW)	6.5	8.4	NA	NA	18.5	10	6	20
FOM (dBc/Hz)	-182	-147	NA	NA	-182.7	-182	-187	-186.63
FOM$_T$ (dBc/Hz)	-198.3	-169.7	NA	NA	-188.3	-194.2	-204.5	-205.58

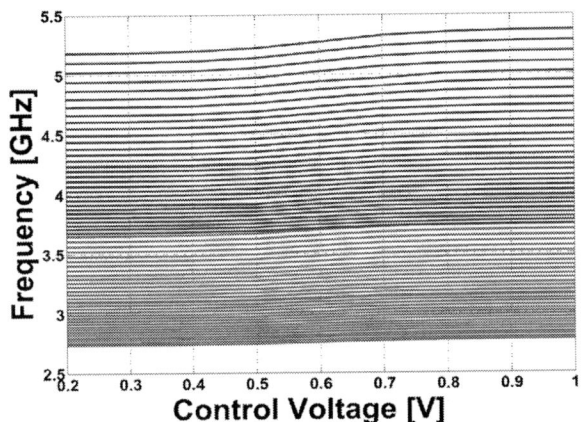

Fig. 4. Measured VCO tuning range.

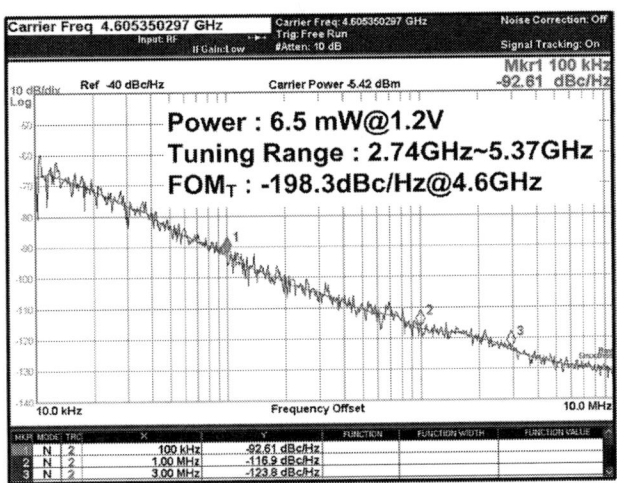

Fig. 5. Measured VCO Phase noise.

synthesizer feedback loop, the final output frequency range covers 50MHz to 5.4GHz. The synthesizer is measured with the reference crystal of 13MHz. A 20b DSM is synthesized to implement fractional-N phase lock loop. Table I shows the performance comparison with other published VCO works, where *FOM$_T$* stands for the figure-of-merit (*FOM*) including the tuning range. The VCO using other technology or the VCO based on the transformer design are also compared after converting the phase noise performance back to 4-5GHz band according to Leesson's equation. The presented dual-path, dual band VCO exhibits the *FOM* of -182dBc/Hz and *FOM$_T$* of -198.3dBc/Hz, which are comparable to the recent VCO works in the literature.

IV. CONCLUSION

The wideband VCO is fabricated in a 130 nm CMOS process and occupies an area of 0.6mm^2. This work demonstrates a low power dual-path, mode switching technique in VCO design. Proposed additional capacitor helps the tank Q optimization in two mode operations. Without phase noise performance penalty, low gain ratio in dual-path architecture improves the immunity to noise coupling in coarse-tuning path. The measured K$_{vco}$ are ~345MHz/V and ~77MHz/V in high band and low band. This VCO consumes 6.5 mW power from a 1.2 V supply and exhibits -123.8dBc/Hz at 3MHz offset from a 4.605GHz carrier showing *FOM$_T$* of -198.3dBc/Hz. The achieved continuous tuning range covers 2.74GHz-5.37GHz by VCO and covers 50MHz-5.4GHz wireless communication bands by synthesizer.

REFERENCES

[1] "Feasibility study for further advancements for E-UTRA (LTEadvanced)", 3GPP, Sophia-Antipolis,France, TR 36.912 Std.

[2] C. e. a. Stevenson, "Ieee 802.22: The first cognitive radio wireless regional area network standard," *Communications Magazine, IEEE*, vol. 47, no. 1, pp. 130–138, January 2009.

[3] T. Rapinoja, "A digital frequency synthesizer for cognitive radio spectrum sensing applications," *Microwave Theory and Techniques, IEEE Transactions on*, vol. 58, no. 5, pp. 1339–1348, May 2010.

[4] K.C. P. et al., "Enhancement of frequency synthesizer operating range using a novel frequency-offset technique for LTE-A and CR applications," *Microwave Theory and Techniques, IEEE Transactions on*, vol. 61, no. 3,

[5] Y. S. et al., "A 2.74 - 5.37ghz boosted-gain type-i pll with < 15 area," in *Radio Frequency Integrated Circuits Symposium (RFIC), 2012 IEEE*, June 2012, pp. 181–184.

[6] W. R. et al., "A uniform bandwidth pll using a continuously tunable single-input dual-path lc vco for 5gb/s pci express gen2 application," in *Solid-State Circuits Conference, 2007. ASSCC '07. IEEE Asian*, Nov 2007, pp. 63–66.

[7] Y. S. et al., "Dual-path lc vco design with partitioned coarse-tuning control in 65 nm cmos," *Microwave and Wireless Components Letters, IEEE*, vol. 20, no. 3, pp. 169–171, March 2010.

[8] J. Z. et al., "A 0.4-6 ghz frequency synthesizer using dual-mode vco for software-defined radio," *Microwave Theory and Techniques, IEEE Transactions on*, vol. 61, no. 2, pp. 848–859, Feb 2013.

[9] L. Fanori, "A 2.4-to-5.3ghz dual-core cmos vco with concentric 8-shaped coils," in *Solid-State Circuits Conference Digest of Technical Papers (ISSCC), 2014 IEEE International*, Feb 2014, pp. 370–371.

[10] M. e. a. Taghivand, "A 3.24-to-8.45ghz low-phase-noise mode-switching oscillator," in *Solid-State Circuits Conference Digest of Technical Papers (ISSCC), 2014 IEEE International*, Feb 2014, pp. 368–369.

Design of Fully Integrated Receiver Front-End for VSAT Applications

Ping-Yi Wang[1], Yun-Chun Shen[1], Min-Chih Chou[1], Yin-Cheng Chang[1,2], Te-Lin Wu[1], Da-Chiang Chang[2], and Shawn S. H. Hsu[1]

[1]Institute of Electronics Engineering, National Tsing Hua University, Hsinchu, Taiwan

[2]National Chip Implementation Center, National Applied Research Laboratories, Hsinchu, Taiwan

Abstract — A fully integrated receiver front-end for very small aperture terminal (VSAT) applications in a 0.18-μm SiGe BiCMOS technology is demonstrated. To satisfy different specifications of various applications, the proposed receiver can down-convert the input signal in a wide RF range from 9.8 to 14.8 GHz to the IF frequency at L-band (950-2150MHz) with four differences LO frequencies. The noise figure is better than 7 dB with an averaged conversion gain of 45.5 dB for the entire RF band. The receiver front-end circuit demonstrate a high linearity (OP1dB > 4.5 dBm) with an excellent gain flatness (±1.5 dB) under a low power consumption (150mW).

Index Terms — SiGe BiCMOS, receiver front-end, LNB, down converter, very small aperture terminal (VSAT).

I. INTRODUCTION

Very-small-aperture-terminals (VSATs) are commonly used for the satellite communication systems. Typical applications are internet access in the remote areas, digital broadcast satellite (DBS), and maritime communications. This paper focuses on the design and implementation of a wideband heterodyne receiver front-end for VSAT applications. In the VSAT systems, the low-noise block (LNB) plays a critical role in determining the signal quality [1]. Considering the requirements of the path loss from satellite to the user-end, the LNB circuits should guarantee a noise figure (*NF*) lower than 0.6 dB and provide a conversion gain up to 60 dB. The LNB circuits are typically realized using the relatively expensive discrete components in III-V technology and combined with the microstrip line filters to meet the specifications [2]. Note that two or three cascaded GaAs HEMT amplifiers are often placed in the front of the LNB circuits to obtain such a low noise and high gain performance, as illustrated in Fig. 1. The rest of the receiver still needs to achieve a noise figure less than 8dB at the entire RF band, which is not an easy task with the Si-based technology. Besides, a conversion gain of 40 dB and the output P_{1dB} compression point of 0 dBm are also the required performance for the rest sections of the receiver.

In this paper, a fully integrated receiver front-end circuit for VSAT application is presented, which consists of a low-noise amplifier (LNA), a mixer, and a broadband IF amplifier, as shown in Fig. 1. Many works were previously reported for Ku-band applications, but only a few studies showed wideband performance using Si-based technologies [3]-[4]. A 10-12 GHz DVB-S receiver in 0.8-

μm silicon bipolar technology was proposed using a two-stage common-emitter LNA with the emitter degeneration inductors technique to achieve low noise performance [4]. However, the design requires many inductors which results in a large chip area and degraded *NF*. Another receiver circuit used the shunt-shunt feedback topology to provide the wideband input power matching [5]. But using the reactive feedback approach exhibits drawbacks of a low power gain with a large dc current. In our previous study, the receiver front-end using the transformer feedback inductor topology provides a moderate RF bandwidth and low noise performance [6]. However, the input impedance of the LNA is mainly determined by the input transistors, and the topology is also sensitive to the parasitic capacitance of the transformer. In this work, the proposed receiver front-end circuit employs the cross-coupled capacitors with transformer inductor as matching networks, which has the advantages of high gain, low noise, and wide bandwidth. As a result, the system requirement can be relaxed from the more expensive discrete pHEMT amplifiers and the overall cost can be reduced.

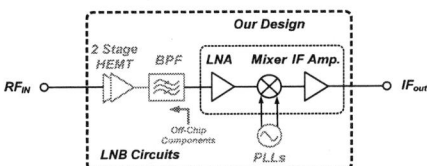

Fig. 1. System block of a wideband reviver front-end circuit.

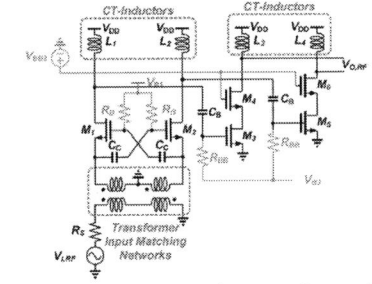

Fig. 2. Proposed two-stage LNA using transformer inductors.

II. CIRCUIT DESIGN

The LNA is a critical block for the system performance, which should provide both power and noise matching in a wide bandwidth in this design. A straightforward approach is using the common-emitter topology with a high-order LC matching networks, but which requires a large chip area. In this study, a two-stage common-gate LNA with cross-coupled capacitors (CCC-LNA) [7] is chosen for reducing the amount of inductors, as shown in Fig. 2. The CCC-LNA topology is suitable for wideband operation due to the inherent property of common-gate topology. The cross-coupled capacitors at the input also help to improve the gain, stability, linearity, and reversed isolation. The CCC-LNA also shows better noise performance for high frequency operation.

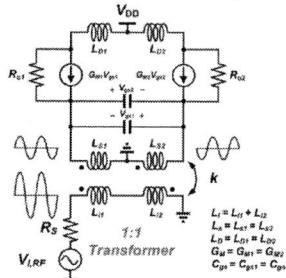

Fig. 3. Incremental model of the proposed LNA.

In order to convert the single-end RF signal to differential, a transformer is used at the input of LNA. The two balanced ports of the transformer–type balun are directly coupled to the source of the input transistors with the center-tapped port grounded which also provides the DC-bias path for the LNA. The input equivalent circuit of the LNA is established to analyze the input matching network, as shown in Fig. 3, where R_s is the resistor of the signal source, M is the mutual inductance between L_i and L_s, C_{gs} and G_M are the gate-to-source capacitance and the transconductance of the input transistors, respectively. Therefore, the input impedance is given by,

$$Z_{in} = sL_i - 2sM \frac{2sM(2sC_{gs} + G_M)}{1 + 2sL_s(2sC_{gs} + G_M)} \quad (1)$$

Assuming that $G_M >> |2sC_{gs}|$, and $G_M = 1/R_s$, Z_{in} can be expressed as

$$Z_{in} \approx \frac{2sL_i[R_s + 2(1 - 2k^2)sL_i]}{R_s + 2sL_i} \quad (2)$$

Based on (2), the input impedance is mainly determined by the input balun (L_i and k) and insensitive to the parasitic capacitance of the input transistors.

In order to reduce the NF and eliminate additional noise contributed from the input transistors, the cross-coupled capacitor topology is used to enhance the transconductance of the amplifier. The transconductance of the input stage can be derived as

$$G_{M,eff} \approx \frac{C_{gs} + 2C_C}{C_{gs} + C_C} G_m \quad (3)$$

And the noise factor can be approximated as

$$F_{CCC} \approx 1 + \frac{\gamma}{\alpha} \frac{C_{gs} + 2C_C}{C_{gs} + C_C} \quad (4)$$

Hence the effective transconductance is increased and noise factor is reduced with the same power dissipation. The analytical equations suggest that the CCC technique offers the advantages of gain, dc power consumption, and NF performances. To make the $1/G_M$ equal to the source impedance R_s, the input transistor with a high V_{GS} and small size should be chosen for wideband matching. To make better trade-off among the NF, power, and chip area, the bias and the transistor size are designed as 0.75 V and 60 μm, respectively. The input transformer is designed as 3 turns with the top metal width of 6 μm. The spacing is designed as 3μm to achieve a coupling factor k of 0.58 with the inner diameter D of 140 μm. The extracted parameters of the transformer are L_i, L_s, n, Q_i, and Q_s of 660 pH, 250 pH, 1.63, 7.64, and 6.42, respectively.

The second stage is the conventional cascode amplifier to enhance the overall gain performance. Compared with the common-source amplifier, the cascode topology offers the advantages of better gain performance, improved stability, and enhanced reverse isolation.

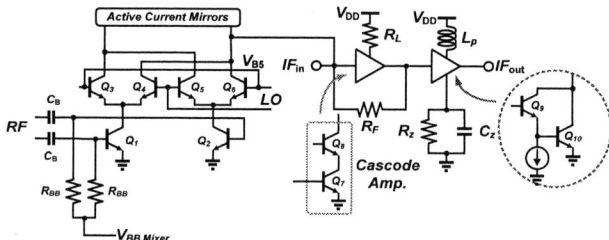

Fig. 4. Circuit blocks of the mixer and IF amplifier.

The circuit topologies of the double-balanced mixer and the IF amplifier are both shown in Fig. 4, which are similar with our previous design [6]. The double-balanced mixer is designed for better port-to-port isolation and linearity, with a moderate gain of about 5 dB. Q_1 and Q_2 act as G_M stage to convert the RF voltage to the current form, and Q_{3-6} act as the mixing stage converter the RF and LO to IF signals. The active current mirror summarized the differential IF signal into the single-end. The IF amplifier is composed of a cascode amplifier for signal amplification, and a Darlington-based output buffer for high output P_{1dB}. The feedback resistor R_F is used to improve the linearity and enhance the signal bandwidth. The emitter degeneration resistor and capacitor create a zero for boosting the high frequency gain.

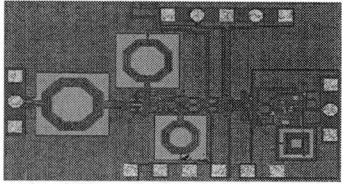

Fig. 5. Photograph of the proposed receiver front-end circuit.

978-1-4799-8198-4/15 $31.00 © 2015 IEEE

TABLE I
PERFORMANCE SUMMARY AND COMPARISON WITH PRIOR WORKS

	Technology (f_T GHz)	Supported Band (GHz)	Gain (dB)	NF (dB)	S_{11} (dB)	OP_{1dB} (dBm)	OIP_3 (dBm)	P_{DC} (mW)	Area (mm²)
[4]	SiGe 0.80 μm (46)	10.7-12.8	38±7	7	<-10	> 5	16	540[b]	N.A.
[5]	SiGe 0.13 μm (200)	8-18	56[a]	6.7-7.8	<-8.5	>-10	N.A.	180	1.81
[6]	SiGe 0.18μm (70)	10.7-13.5	51±1	5-5.8	<-10	> 5.5	> 18	135	0.60
[8]	SiGe 0.25 μm (110)	10.7-12.8	43±1	6-7	<-10	7	16	130/260[c]	1
This work	**SiGe 0.18 μm (70)**	**9.8-14.8**	**45±1.5**	**6.2-7**	**<-10**	**> 4.5**	**> 16**	**150**	**0.86**

(a): include variable-gain amplifier. (b): include PLL. (c): include PLL

III. MEASURED RESULTS AND DISCUSSION

The receiver front-end IC was implemented in the standard TSMC 1P6M 0.18-μm BiCMOS process. The chip photograph is shown in Fig. 5 with a chip size of 0.86 mm², including the RF and dc bias pads. The LNA consumes 9 mA from a 1.8 V supply voltage, and the receiver front-end circuit consumes 35 mA with the total power consumption of 150 mW.

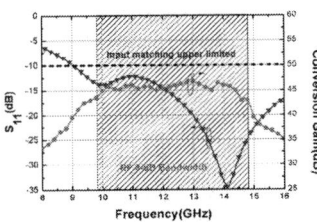

Fig. 6. Measured S_{11} and RF bandwidth of the receiver front-end.

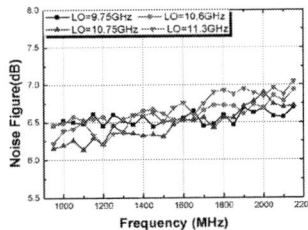

Fig. 7. Measured NF versus IF frequency at four LO frequencies.

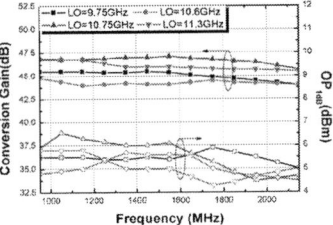

Fig.8. Measured conversion gain and OP_{1dB} versus IF frequency at four LO frequencies.

Fig. 6 shows the RF input return loss and RF bandwidth of the receiver front-end. The input matching achieves better than -10 dB over the Ku-band range, with a RF 3-dB bandwidth about 5 GHz. Fig. 7 shows the measured NF performance of the receiver, which is between 6.2 to 7 dB in the L-band range. The average measured conversion gain is about 45.5 dB with a gain flatness within ±1.5 dB and output P_{1dB} compression point

above 4.5 dBm in the entire band, as shown in Fig. 8. The measured performance and the comparison with previous works are summarized in Table I.

IV. CONCLUSION

This work successfully demonstrated a wideband receiver front-end in a low cost 0.18-μm SiGe BiCMOS technology. The proposed receiver front-end takes the advantage of using cross-coupled capacitors in the common-gate LNA, which could enhance the overall noise performance and the RF 3-dB bandwidth. The proposed receiver front-end is suitable for the application in high performance VSAT systems.

ACKNOWLEDGEMENT

The authors would like to thank the Chip Implementation Center (CIC), Hsinchu, Taiwan for chip implementation and high frequency measurement support by Shawn-Guann Lin and Ya-Wen Ou.

REFERENCES

[1] Gerben W. de Jong, *et al.*, "A fully integrated Ka-band VSAT down-converter," *IEEE J. Solid-State Circuits*, vol. 48, no.7, pp. 1651-1658, Jul. 2013

[2] A. Mason, "A digital video broadcasting standards for satellite, terrestrial and cable television transmission," in *IEEE Microw. Symp. Dig.*, 1998, pp.61-66.

[3] Z. Deng, *et al.*, "A CMOS Ku-band single conversion low-noise block front-end for satellite receivers," *in Proc. RFIC, 2009*, pp. 135–138.

[4] G. Girlando, *et al.*, "A monolithic 12-GHz heterodyne receiver for DVS-S application in silicon bipolar technology," *IEEE Trans. Microw. Theory Tech.*, vol. 53, no. 3, pp. 952-959, Mar. 2005.

[5] D. Ma, *et al.*, "An X-and Ku-band wideband recursive receiver MMIC with gain-reuse," *IEEE J. Solid-State Circuits*, vol. 46, pp. 562-571, Mar. 2011.

[6] Ping-Yi Wan, *et al.*, "A fully integrated Ku-band down-converter front-end for DBS receivers," in *IEEE Microw. Symp. Dig.*, 2014.

[7] W. Huo, *et al.*, "A capacitor cross-coupled common-gate low-noise amplifier," *IEEE Trans. Circuits Syst. II, Exp. Briefs*, vol. 52, no. 12, pp. 857–879, Dec. 205.

[8] P. Philippe, *et al.*, "A low power 9.75/10.6 GHz down-converter IC in SiGe:C BiCMOS for Ku-Band satellite LNBs," *in Proc. BCTM 2011*, pp. 211–214.

Review of Silicon-based Millimeter-wave Radio Frequeny Integrated Circuits

(Invited Paper)

Huei Wang

Dept. of Electrical Engineering and Graduate Institute of Communication Engineering,
National Taiwan University, Taipei, Taiwan, 10617, ROC

SUMMARY

III-V based MMICs such as GaAs and InP were popular for millimeter-wave (MMW) applications at early time due to their superior performances. After the year 2000, silicon-based technologies gradually took over GaAs-based technologies in popularity for some III-V niche applications. Silicon-based technologies have been advanced aggressively and rapidly. Based on Moore's law, the period for a doubling in chip performance, including a combination of the effect of more transistors and their higher speed, is 18 months. Thus, silicon-based technologies are promising and their performances can predictably catch up with those GaAs-based technologies for some aspects in the system applications. Also, silicon-based technologies have much better integration capability than GaAs-based technologies. It is easier to integrate the baseband digital circuits into a single chip with MMW circuits in silicon-based technologies rather than that in GaAs-based technologies, which is a very important feature in many products.

In this talk, the advances of the silicon-based millimeter-wave (MMW) radio frequency integrated circuits (RFICs) will be presented. The silicon-based technologies for MMW RFICs will be introduced briefly. In addition, the current status of the MMW RFICs will be surveyed and novel circuit topologies are summarized. Some representative MMW RFICs are illustrated as the design examples in the categories of their functions in a MMW system. Finally the future trend of the development of the MMW ICs will be addressed.

A 122-150 GHz LNA with 30 dB Gain and 6.2 dB Noise Figure in SiGe BiCMOS Technology

Roee Ben Yishay, Evgeny Shumaker and Danny Elad

IBM Haifa Research Lab, Mount Carmel 31905 Haifa, Israel

Abstract — This paper presents a D-Band Low Noise Amplifier (LNA) for imaging applications designed in an advanced 90 nm SiGe BiCMOS process. The LNA consists of 3 cascode stages and one common-emitter stage and achieves peak gain of 30 dB and 3 dB bandwidth at 122-150 GHz. The measured noise figure is maintained below 7.5 dB over the measurement band with minimum of 6.2 dB at 140 GHz. The amplifier consumes DC power of only 45 mW, and occupies area of 0.11 mm^2 (without pads), which make it suitable for integration in large scale imaging arrays. To our knowledge, this work demonstrates the highest gain, highest bandwidth and lowest noise figure achieved so far at D-Band in any silicon-based technology.

Index Terms — low noise amplifier, SiGe, D-Band, millimeter wave integrated circuits.

I. INTRODUCTION

Millimeter wave systems operating in the D-Band frequency range (110-170 GHz) have a potential use in high resolution imaging due to the atmospheric attenuation window around 140 GHz. Passive direct detection systems relies on the sensing of black-body radiation from an object. Such imagers typically require a high gain, wideband low noise amplifier to amplify the signal to the detector while avoiding from becoming a bottleneck on the system's minimum achievable thermal resolution (NETD). SiGe BiCMOS platform is capable of providing high yield needed for large scale arrays that integrate mm-Wave components and readout circuitry on the same chip and thereby reducing assembly cost. Moreover, recent advances in SiGe technology result in HBT devices with cutoff frequencies exceeding 300 GHz which enable to realize high gain amplifiers well above 100 GHz [1]-[3] with low DC power and area consumption which are critical factors for large arrays.

II. TECHNOLOGY

The LNA was fabricated in IBM's 90 nm SiGe BiCMOS technology (IBM9HP). The process features high-speed SiGe HBT transistors with f_{MAX}/f_T=350/300 GHz and high performance Shottcky Barrier and PIN diodes. The process offers 10 metal stack with one thick (3 μm) copper layer and one thick (4 μm) aluminum layer which can be used for low loss interconnects. The BEOL passives include 2 types of MIM capacitors (high-Q and high density), as well as TaN resistors. CMOS transistors are also available for digital and analog circuitry.

III. CIRCUIT DESIGN

The prevalent topology choice for LNA design in mm-Wave frequencies is the common-emitter [4], [5], owing to its low noise figure and simple matching. However, at D-Band the HBT suffers from low gain due to miller multiplication of C_{be}, which enhance noise contribution from subsequent stages. For example, a 4 μm CE device biased at 4 mA (between NF_{min} and MAG current densities) exhibits MAG and NF_{min} of 5.5 dB and 5.2 dB, respectively, simulated at 140 GHz after parasitic extraction. The cascode configuration, on the other hand, exhibits much higher gain (MAG of 12.5 dB for the same sizing and current density), but with NF_{min} higher by 1dB. Assuming the insertion loss of the interstage matching networks is 1.5 dB, three stages CE amplifier has higher accumulated noise figure and lower gain than two cascode stages amplifier. Therefore, the LNA was designed using 3 cascode stages operating with 3 V supply and a single CE as the output stage, operating with 1.5 V supply, as shown in Fig. 1. The CE stage facilitates wideband matching, which is essential for efficient power transfer to the power detector (not shown here), which also uses the same supply. Device sizing and biasing were determined in 2 steps. First, a sweep of the emitter length and collector current was done to find the current density (J_C) that yields a good compromise between noise figure and gain at 140 GHz (0.85-0.9 mA/μm). Then, the feasibility of wideband matching is considered by evaluating the distance of input impedance (Γ_{in}) and optimal noise source impedance (Γ_{opt}) from 50 Ω. In a sweep on the device size from 2 μm to 8 μm (Fig. 2), a 4 μm HBT was found optimal in that aspect, while satisfying power consumption constrain. The output CE stage has 2 μm long emitter with higher current density (2 mA), to minimize parasitic capacitance and maximize the gain while setting the output impedance as close to 50Ω as possible. One of the design objectives was to achieve high bandwidth (>25 GHz). For that aim, the intersatge matching networks are designed in a staggered fashion. Matching networks design relies on shielded 50 Ω microstrip transmission lines (realized between the thick Cu layer as signal and the first 5 metal layers as ground),

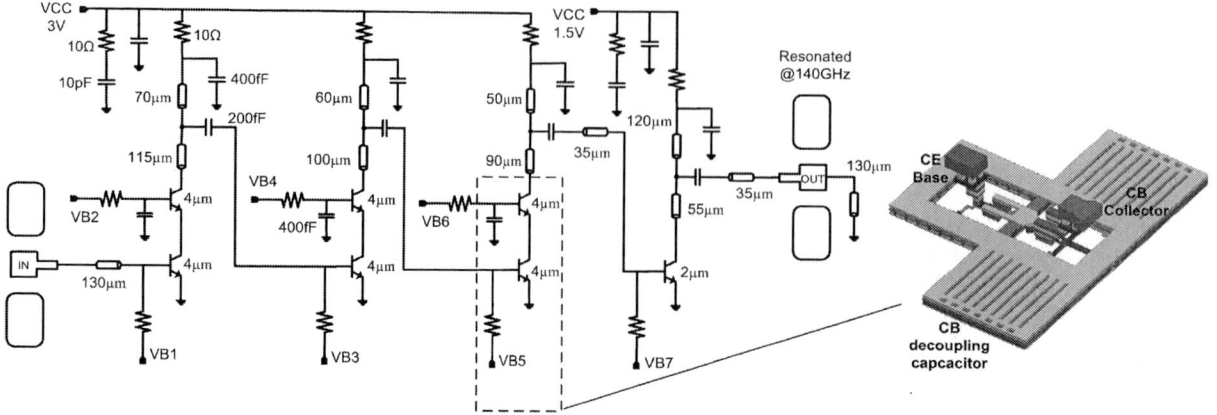

Fig. 1. Schematic diagram of the LNA with 3D layout view of the cascode cell (all T-lines with $Z_0=50\Omega$)

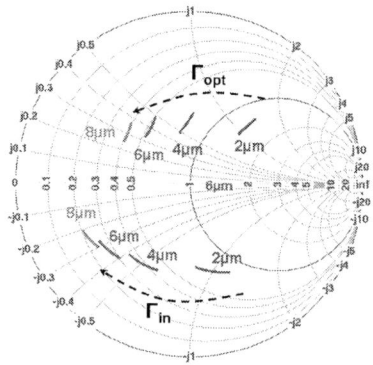

Fig. 2. Simulated Γ_{opt} and Γ_{in} for 2-8μm HBT with Jc=0.3-1.3mA/μm (f=140 GHz)

along with high density MIM capacitors. The design makes an extensive use of interconnects (T-lines, bends, T-junctions) models and parametric cells provided by the design kit. All MIM capacitors are large enough (200 fF at low frequency) to act merely as DC blocks, to minimize the design sensitivity to their exact value due to self resonance effect. The decupling for the supply nodes and the bases of the common-base transistors is realized using custom slotted plates MOM capacitors implemented with the BEOL lower level metals (M1 to M8 and M1 to M5, respectively).

The transistors' contact structure up to the wiring level to transmission lines (illustrated in Fig. 1) was modeled using Agilent Momentum EM solver to capture all high frequency effects, such as parasitic inductances of vias and traces, as well as their mutual magnetic coupling, which is absent from RC extraction. Input and output pads are also EM-characterized since the input pad is a part of the input matching network, while the output pad is resonated by a short stub at 140 GHz to enable to present 50 Ω load to the circuit without needing any de-embedding.

The LNA was designed for unconditional stability at all frequencies. At low frequency, the high gain of the devices is suppressed by series 10 Ω resistors placed on the supply path and on the common base nodes (after the decoupling capacitors). Further stability enhancement in intermediate frequencies is achieved by the RC sections tied between supply and ground, as well as supply routing in M4, which acts as high loss stripline above 10 GHz.

IV. MEASUREMENT RESULTS

The chip photograph of the fabricated LNA is shown in Fig. 3. The IC occupies an area of 330 μm × 350 μm without pads and 700 μm × 750 μm including pads. It consumes 14 mA from 3 V supply and 2 mA from 1.5 V supply (45 mW). S-Parameters were measured using Agilent N5227A PNA with OML Inc. D-Band extension modules. The result presented in Fig. 4 after SOLT probe tip calibration. The LNA achieves peak gain of 30 dB, and 3 dB bandwidth extended from 122 to 150 GHz with good correlation to simulated data. The measured input return loss remains below -10 dB in the entire frequency range, while S22<-10 dB at 130-150 GHz.

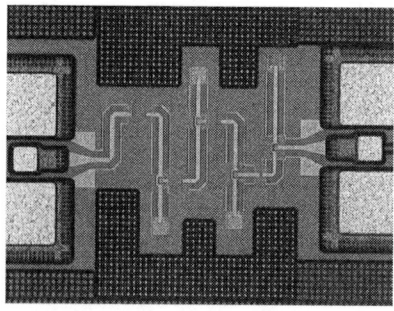

Fig. 3. LNA chip photograph

978-1-4799-8198-4/15 $31.00 © 2015 IEEE

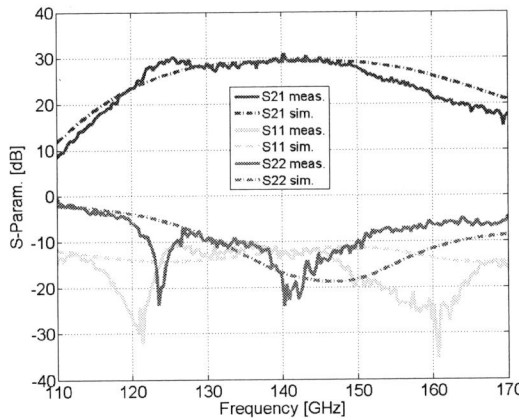

Fig. 4. Measured and Simulated S-Parameters

Noise figure measurement was done using ELVA-1 D-Band noise source at the input and VDI sub-harmonic mixer (CL=5-6 dB) at the output to downconvert the signal to a fixed 1 GHz IF, analyzed by Agilent E4448A PSA. The 60-75 GHz LO signal to the mixer is supplied by a Hittite T2240 signal generator followed by SAGE ×3 frequency multiplier. The insertion loss of GGB WR-6 probes and waveguide parts was verified individually and de-embedded from the measurement. A minimum averaged noise figure of 6.2 dB is obtained around 140 GHz, and maintained below 7.5 dB at 120-150 GHz, as shown in Fig. 5. The measured noise figure is up to 0.7 dB lower than simulated, but follows the same trend. This deviation is consisted with the results in [5], which was based on same technology and fabricated in the same run. The gain measurement obtained from the noise figure setup is in good agreement with the gain measured in the independent S-parameters setup.

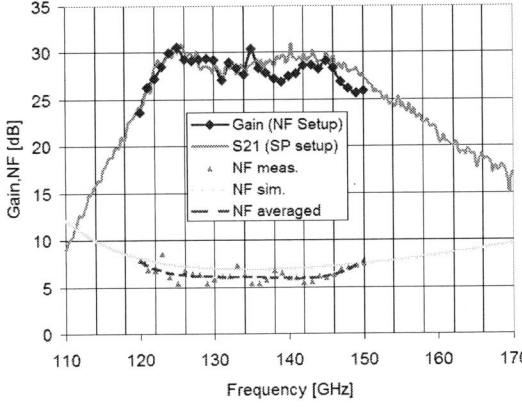

Fig. 5. Measured and Simulated Noise Figure and Gain

TABLE I
RESULTS SUMMARY AND PERFORMANCE COMPARISON

	[1]	[2]	[3]	This
Technology f_T/f_{max} [GHz]	260/350	285/330	300/500	300/350
Topology	Diff 3xCC	2xCE+ 3xCC	2xCC	3xCC+ 1xCE
Freq. [GHz]	162	165	110	140
Gain [dB]	24	31	20	30
BW [GHz]	12	7	20	28
NF [dB]	7.4 (**)	11(*)	4	6.2
P_{DC} [mW]	60	92	17	45
Area [mm²]	0.06	0.074	0.41	0.11

(*) simulated   (**) without balun

V. CONCLUSION

A state of the art D-Band LNA implemented in a 90 nm SiGe BiCMOS technology was presented. A comparison of this LNA to other recently published work is shown in Table 1. To our best knowledge, this is the first reported LNA with NF below 7 dB at D-Band in a silicon based technology, along with the best gain-bandwidth product. Its small footprint and power consumption make it suitable for integration in a large scale passive imaging array.

VI. ACKNOWLEDGMENT

The authors would like to thank D. Harame, J. Pekaric and the IBM 9HP team for their support in this work.

REFERENCES

[1] E. Öjefors, F. Pourchon, P. Chevalier and U.R. Pfeiffer, "A 160-GHz Low Noise Downconverter in a SiGe HBT Technology," *European Microwave Week Conf.*, Sep 2010.

[2] E. Dacquay, A. Tomkins, K. Yau, E. Laskin, P. Chevalier, A. Chantre, B. Sautreuil, and S. P. Voinigescu, "D-Band Total Power Radiometer Performance Optimization in an SiGe HBT Technology," *IEEE Trans. Microwave Theory and Tech.* vol. 60, no. 3, pp. 813-826, Mar. *2012*

[3] A. Ulusoy, M. Kaynak , V. Valenta, , B. Tillack and H. Schumacher, "A 110 GHz LNA with 20 dB Gain and 4 dB Noise Figure in an 0.13 µm SiGe BiCMOS Technology," *IEEE Int. Microwave Symp.* ,June 2013.

[4] O. Katz, R. Ben Yishay, R. Carmon, B. Sheinman, D. and D. Elad, "High-Power High-Linearity SiGe-Based E-Band Transceiver Chipset for Broadband Communication," *IEEE Radio-Frequency Integrated Circuits Symp.* , June 2012.

[5] P. Song, A. Ulusoy, R. L. Schmid and J. D. Cressler, "A High Gain, W-Band SiGe LNA with sub-4dB Noise Figure," *IEEE Int. Microwave Symp.*, June 2014.

120GHz Low Power, High Gain, Wideband Active Balun For Chip-to-Chip Communication

Chae Jun Lee, Hae Jin Lee, Dongmin Kang, In Sang Song, Hong Yi Kim, Seong Jun Cho,
Joong Geun Lee, In- Yeal Oh, and Chul Soon Park

Department of Electrical Engineering KAIST, 291 Daehak-ro, Yuesong-gu, Daejun 305-701,Korea
chaejunlee@kaist.ac.kr

Abstract — This paper presents a 120 GHz low power, high gain, wideband active balun design in 65nm CMOS. The active balun is realized using current-reuse cascode topology and common source topology. The active balun exhibits a measured small signal where S21 and S31 are -5 ± 1.3 dB and -4.8 ± 0.5 dB, respectively, from 113GHz to 133 GHz. The measured gain imbalance and phase imbalance is kept less than 1.5 dB and 2°, respectively, from 113 GHz to 133 GHz. The chip occupies 460 × 460 μm² including the pad. Total power consumption is 4mW from a 1 V supply voltage. To our best knowledge, this is the first active balun operating in the D-band.

Index Terms — Active balun, CMOS, D-band, low power., Millimeter-wave

I. INTRODUCTION

Recently, high-data rate communication systems using millimeter waves have been emerging, such as 60 GHz wireless systems. In the 60 GHz band, many systems with a high-speed, low cost have been reported [1]-[3]. These systems have only a 7 GHz bandwidth, which limits high-data rate communications of over several Gbps. As the electronics industry develops, there is a big demand for higher bandwidth communication for applications, such as 3D video (3-Gbit/s) and 4K digital movies (6-Gbit/s). Inevitably, to meet this demand for bandwidth, systems operating at sub-THz are necessary because those systems provide sufficient bandwidth. NTT has reported on MMICs [4] for 10-Gbit/s data transmission in the sub-THz region. One promising high-speed system is chip-to-chip communication.

Many systems use differential topologies to overcome disadvantages inherent in single-ended circuits, for example, crosstalk cancellation, noise and common-mode rejection. Despite having advantages, some blocks such as antennas and low noise amplifiers have been designed as single-ended to ensure a small size and low power application. Consequently, the balun is very crucial in receiver front-ends to convert a single-ended signal to a differential signal.

Baluns can be categorized into passive baluns and active baluns. Passive baluns usually suffer from having a larger

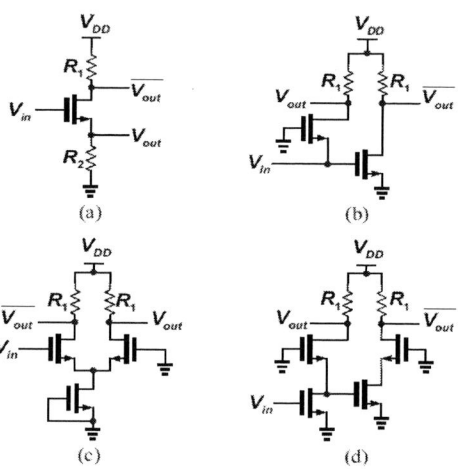

Fig. 1.   Conventional active balun circuits.

area and higher insertion loss over 100 GHz. It is hard to achieve high gain performance at sub-THz frequency due to the limitation of transistors' fmax and ft in the CMOS process. Active baluns, which offer gain are a great alternative solution, though active devices have the weakness of power consumption. Fig. 1 [5] shows several conventional active baluns. Circuit topologies are quite simple and the difference of phase is relatively accurate. Nevertheless, these topologies have problems with a large gain imbalance as the operating frequency increases.

In this paper, a 120 GHz low-power active balun is designed adopting current-reuse cascode and common source. The measured results show that the active balun has small signal gains of -5 dB ± 1.3 dB (S21) and -4.8 ± 0.5 dB (S31) from 113 GHz to 133 GHz. The amplitude imbalance in the desired band is less than 1.5 dB and the phase imbalance is 2°. The active balun consumes 4mW DC power and the chip area is 460 × 460 μm² including all pads.

II. CIRCUIT DESIGN

The active balun is fabricated using the TSMC 65nm CMOS process. Fig. 2 shows a schematic of proposed 120

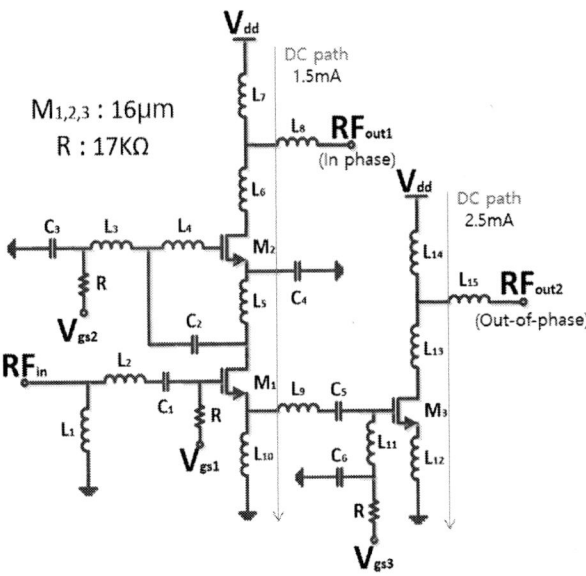

M$_{1,2,3}$: 16μm

R : 17KΩ

Fig. 2.   Schematic of the proposed 120 GHz active balun.

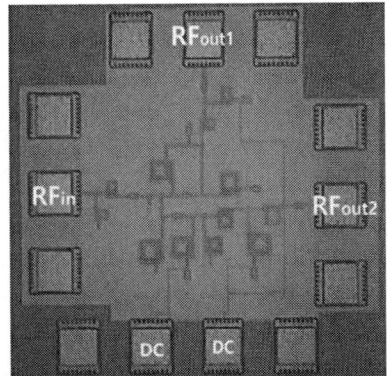

Fig. 3.   Microphotograph of the proposed active balun.

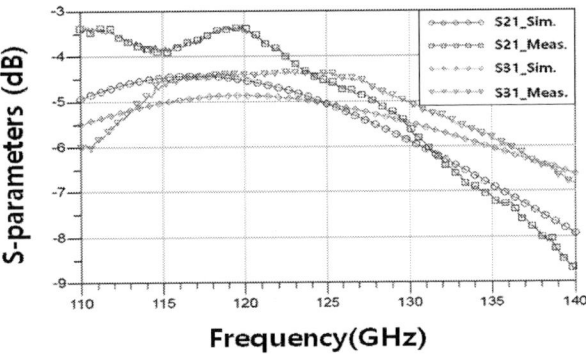

Fig. 4.   Measured and Simulated gain of the proposed 120 GHz active balun.

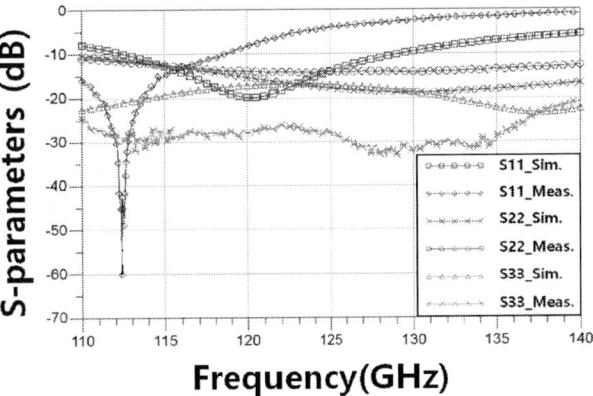

Fig. 5.   Measured and simulated return losses of the proposed 120 GHz active balun.

GHz active balun. The active balun employs a current-reuse cascaded common source topology, it builds an in-phase signal because the input signal is inverted twice, while the out-of-phase signal was made by adopting a common source, since the input signal is inverted once. A current-reuse topology was applied between two cascade common source topology to save DC current consumption. By means of the current-reuse topology, similar gains can be obtained at the in-phase port and out-of-phase port, even though different currents are used at each DC path.

C1, C2, and C5 were employed to block DC current and couple the AC signal. M2 operate as a common source amplifier with the help of C4 which separates the AC paths from the single DC path. C3 and C6 were used as the AC ground, which affects the inter-stage matching network. Capacitors are set to the value seen as the AC ground. Inductors are employed in the role of input matching, output matching, inter-stage matching, as well as for the loads of transistors M2 and M3. The S-parameters of all inductors and lines were decided by EM simulation with Ansoft HFSS and Agilent Momentum. All inductors and lines have a 2.06 μm width. The gate width of M1, M2 and M3 is 16 μm (1 μm × 16 fingers).

III. MEASUREMENT RESULT

The fabricated balun chip is shown in Fig. 3. The 120 GHz active balun, including all pads, has a chip size of 460 × 460 μm². Total power consumption is 4 mW from a 1 V supply voltage. The S-parameter as well as the phase measurement results were obtained from 110 GHz to 170

GHz by using the Agilent PNA with a frequency extender and WR-06 waveguide probes. The input was set to port 1, whereas the in-phase output and out-of-phase port were set to port 2 & 3, respectively. The measured and simulated S-parameters are plotted in Fig. 4 and Fig. 5, respectively. S21 and S31 are -5 dB ± 1.3 dB and -4.8 ± 0.5 dB from 113 GHz to 133 GHz. The input return loss, S11 is less

978-1-4799-8198-4/15 $31.00 © 2015 IEEE

TABLE I
COMPARISON OF STATE-OF-THE-ART ACTIVE BALUNS

Ref.	Technology	Bandwidth (GHz)	Gain (dB)	Gain Imbalance (dB)	Phase Imbalance (Degree)	P$_{DC}$ (mW)	Area (mm²)
[6]	0.13 µm SiGe	54-59	-1.4	1.2	5	10.4	0.63
[7]	0.13 µm CMOS	2-40	1	0.5	10	40	0.56
[8]	0.13 µm SiGe BiCMOS	31-65	12.4	0.5	2.5	278	0.47
[9]	0.25 µm SiGe	0.2-22	5	0.5	4	166	0.7
[10]	0.35 µm SiGe HBT	0.1-70	2.5	2.5	7-17	29.7	0.5
This Work	65 nm CMOS	113-133	-5	1.5	2	4	0.21

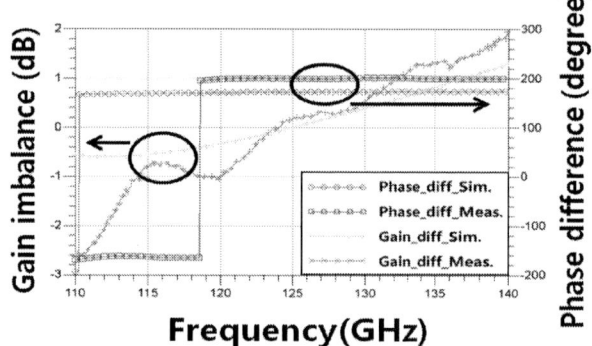

Fig. 6. Measured and simulated gain imbalance and phase difference of the proposed 120 GHz active balun.

than -10 dB from under 110 GHz to 120 GHz. The S11 result is down-shifted about 7 GHz compared to the simulation result. The output return loss, S22 and S33 are less than -10 dB from under 110GHz to over 140 GHz. Fig. 6 illustrates the gain imbalance and phase difference between the in-phase port and out-of-phase port. The gain imbalance is kept less than 1.5 dB from 113 GHz to 133 GHz. The measured phase difference between the in-phase port and out-phase-port is about 200° because the 3-port calibration for phase in the D-band is not accurate. However, the phase imbalance is less than 2°, a desirable value. The measurement results for the S-parameter and gain imbalance are in good agreement with simulation data. Table 1 summarizes state of the art baluns, compared with our proposed active balun.

IV. CONCLUSION

A low power, wideband active balun is designed and fabricated using the TSMC 65nm CMOS process. Across the 113-133 GHz frequency range, the active balun exhibits measured gains of -5 dB ± 1.3 dB and -4.8 ± 0.5 dB at the differential ports. The gain and phase imbalance are 1.7 dB and 2° from 113 GHz to 133 GHz. The total power consumption is 4 mW with a 1 V supply voltage and the chip area of 460 × 460 µm² is small. To our best

knowledge, this is the first active balun for the 120 GHz frequency and it has wideband characteristics while operating with very low power.

ACKNOWLEDGEMENT

This research was funded by the MSIP (Ministry of Science, ICT & Future Planning), Korea in the ICT R & D program 2014

REFERENCES

[1] T. Takayuki et al., "A Fully Integrated 60 GHz CMOS Transceiver Chipset Based on WiGig/IEEE802.11ad with Built-In Self Calibration for Mobile Application." IEEE Intl. Solid-State Circuits Conf. (ISSCC), pp.230-231, Feb. 2013.

[2] K. Okada et al., "A 64-QAM 60 GHz CMOS Transceiver with 4-Channel Bonding," IEEE Intl. Solid-State Circuits Conf. (ISSCC), pp.346-347, Feb. 2014.

[3] K. Lingkai et al., "A 50mW-TX 65mW-RX 60 GHz 4-Element Phased-Array Transceiver with Integrated Antennas in 65nm CMOS," IEEE Intl. Solid-State Circuits Conf. (ISSCC), pp.234-235, Feb. 2013.

[4] T. Kosugi et al., "120-GHz Tx/Rx waveguide modules for 10-Gbit/s wireless link system," in IEEE CSICS., Nov. 2006, pp.25-28

[5] C. S. Lee et al., "A low noise amplifier for a multi-band and multi-mode handset," in IEEE RFIC Symp., Jun. 1998, pp. 47-50

[6] Y. Jin et al., "A 60 GHz-band millimeter-wave active balun with +-5° phase error," in European Microw. Conf., 2010, pp.210-213.

[7] B.-J. Huang et al., "A 2-40 GHz active balun using 0.13 µm CMOS process," IEEE Microw. Wireless Compon. Lett., vol. 19, no. 3, pp. 164-166, 2009.

[8] A. Awny et al., "Broadband 31-65 GHz inductorless active balun with 12.4 dB gain in 0.13µm SiGe:C BiCMOS technology, in European Microw. Conf., 2011, pp. 652-655

[9] C. Viallon et al., "Design of an original k-band active balun with improved broadband balanced behavior," IEEE Microw. Wireless Compon. Lett., vol. 15, no. 4, pp. 280-282, 2005

[10] A. Gharib. et al., "A 70 GHz bandwidth low-power active balun employing common-collector resistive feedback in 0.35 µm bipolar SiGe technology," in IEEE International Microwave Symposium, 2014, pp.1-3

Electronic THz Transmissive Imaging System

Wei-Cheng Chen[1], Chih-Wei Lai[1], Tzu-Chao Yan[1], Chun-Hsing Li[2], Tzu-Yuan Chao[1], and Chien-Nan Kuo[1]

[1]Department of Electronics Engineering, National Chiao-Tung University, Hsinchu, Taiwan

[2]Department of Electrical Engineering, National Central University, Jhongli City, Taiwan

Abstract — An electronic THz transmissive imaging system is presented in this work. This system is composed of a CMOS signal source, a CMOS signal sensor, lenses, and commercial electronic chopping components. The output frequency of the signal source is 339 GHz, and the responsivity of the signal sensor is 200 kV/W. The total dc power consumption of those CMOS circuits is 91.02 mW. Using electronic chopping at the frequency up to 100 kHz helps improve the system dynamic range and makes imaging feasible. The resolution of the realized imaging system achieves 2 mm.

Index Terms — THz imaging, THz source, THz sensor, lens, chopping.

I. INTRODUCTION

Imaging in the THz frequency regime is an emerging research topic in recent years. Due to the properties of transparency in non-metallic materials and specific spectral fingerprint in many materials, the material recognition by the THz imaging has a great potential in different fields. The applications include the security check, the medical imaging, and the agricultural imaging.

Fig. 1(a) shows a typical setup of the THz transmissive imaging system [1]. The signal source and the signal sensor are placed at the focal point of the lenses for maximum power transmission and sensing. The distance between the image object and two lenses are equal to the focal length for power focusing. For weak signal detection, the transmitted signal is modulated by the mechanical chopper and demodulated by the lock-in amplifier. The higher chopper frequency could reduce more noise effect of the output signal [2]. The resolution of the acquired image would be increased. However, the chopper frequency is limited by the rotating speed of the fan, typically smaller than 1 kHz.

For higher chopper frequency requirement, the THz carrier is amplitude modulated by the electronic controlled signal [2]-[4]. The imaging information of these THz detectors is also demodulated by the lock-in amplifier. Nevertheless, the more compact solution for the chopping signal demodulation is more suitable to mobile purpose.

In this work, an electronic THz transmissive imaging system is presented. The signal source and the signal sensor are the customized CMOS integrated circuits in 40 nm and 0.18 μm technology, respectively. The chopping

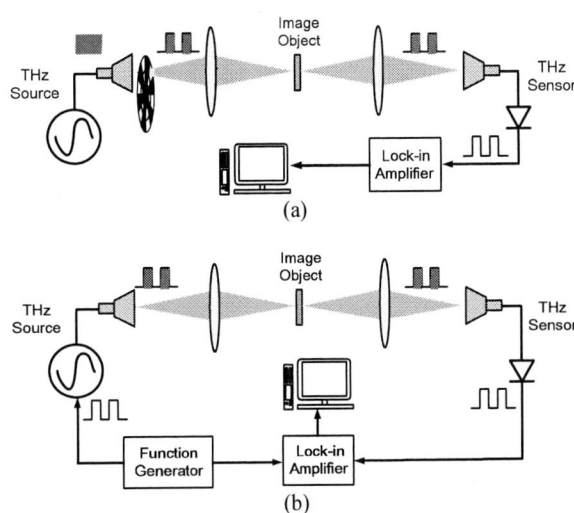

Fig. 1. Block diagram of the THz transmissive imaging system using (a) mechanical chopping and (b) electronic chopping.

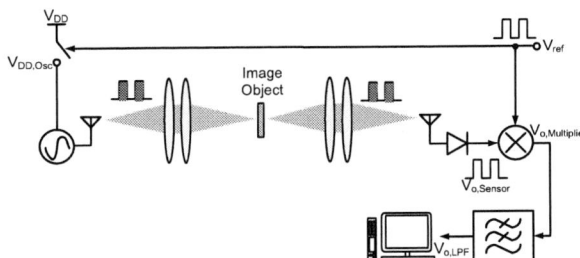

Fig. 2. Block diagram of proposed electronic THz transmissive imaging system.

modulation and demodulation are implemented by the commercial integrated circuits. In brief, a prototype solution for high resolution imaging application is integrated and demonstrated.

II. SYSTEM DESIGN AND IMPLEMENTATION

The block diagram of the proposed electronic THz imaging system is as shown in Fig. 2. This system is composed of a signal source, two lens modules, a signal sensor, a SPST (single-pole, single through) switch, an analog multiplier and a low-pass filter. In this setup, the focal length of the lenses is 10 cm. Both the signal source and signal sensor are placed at the focal point.

The integration of the designed CMOS integrated circuits and the commercial components are described as follows.

A. Signal Source and Chopping Modulation

The published works have exposed different approaches for electronic chopping modulation or OOK (on-off keying) modulation with THz carrier. In [5], a SPST switch received 245 GHz LO carrier and was modulated by the 40 Gbits/sec input data. The insertion loss and isolation are -4 dB and -16 dB, respectively. However, this method would induce extra signal loss, and the signal-to-noise ratio of the system was degraded.

In this work, the signal source chip is a 339 GHz triple-push oscillator implemented in 40 nm CMOS technology [6]. With 1 V supply voltage, the simulated oscillation startup time is less than 200 psec. The simulated transient response of the fundamental frequency at 113 GHz is shown in Fig. 3. From this result, the theoretical ON/OFF switching speed of supply voltage for stable oscillation could be large than hundreds MHz.

The commercial SPST switch (Analog Device ADG801) is employed to control the supply voltage of the CMOS signal source for chopping modulation. The control signal is provided by the function generator. Consequently, the signal source module is realized.

B. Signal Sensor and Chopping Demodulation

Weak signal detection by the lock-in amplifier is multiplying the output signal of the signal sensor and the reference signal. Because these signals are located at the same frequency, the output dc term is proportioned to the magnitude of input signals [7].

In this work, the common-source based THz sensor in 0.18 μm CMOS is used for signal power detection [8]. The responsivity of this THz sensor is 200 kV/W at 339 GHz. The function of lock-in amplifier is implemented by an analog multiplier (Analog Device AD633).

A LPF (low-pass filter) follows the analog multiplier to filter out the non-desired ac signal. The LPF is composed of a shunted 100 kΩ resistor and a shunted 1 μF capacitor. The output dc voltage is recorded by the computer. The value of the dc voltage depends on the input power level and gives the image information.

III. MEASUREMENT RESULT

Fig. 4 shows the photo of the realized electronic THz transmissive imaging system. The PCB (print circuit board) photo of the SPST switch, the signal source, the signal sensor and the analog multiplier are shown in Fig. 5.

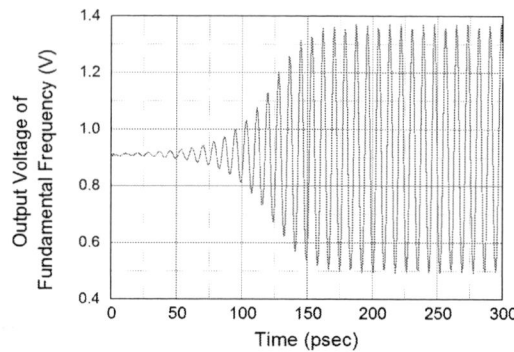
Fig. 3 Transient response of the triple-push oscillator.

Fig. 4. Setup of the proposed electronic THz transmissive imaging system.

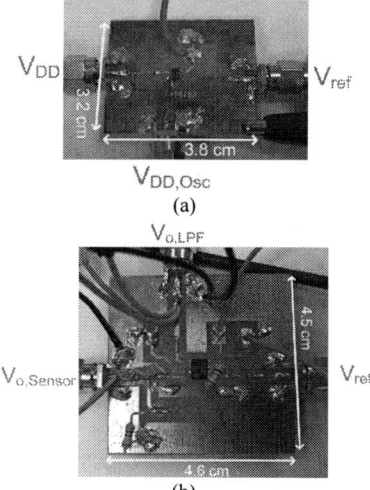
(b)
Fig.5 PCB of (a) the SPST switch, (b) the analog multiplier, and the LPF.

Fig. 6 shows the measurement result of the chopping modulation of signal source. With 100 kHz chopping frequency, the supply voltage of the oscillator would drop down to 0 V which leads the oscillator to the power-off state. The desired output signal for the THz imaging system is generated. The dc power consumption of the CMOS THz source is 37.5 mW from 1 V supply voltage.

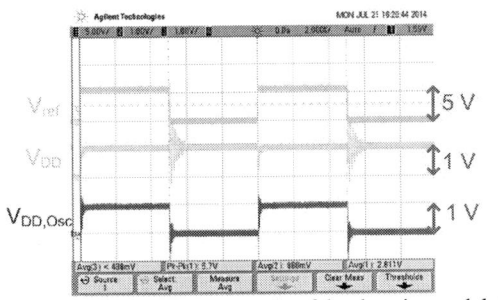

Fig. 6. Time domain measurement results of the chopping modulation.

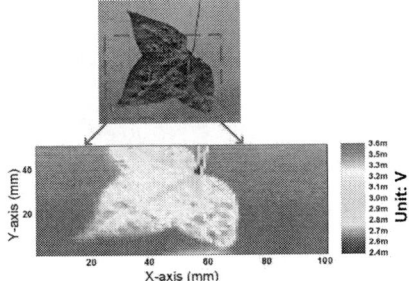

Fig. 7. A leaf for the water content testing.

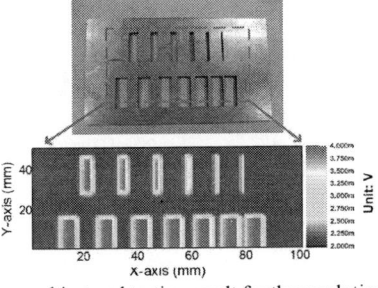

Fig. 8. The image object and testing result for the resolution measurement.

The supply voltage of the CMOS THz sensor and the commercial analog multiplier is 1.8 V and 11.4 V, respectively. The dc power dissipations are 7.92 mW and 45.6 mW, respectively.

In Fig. 7, a leaf is used as an image object. The output dc voltage indicates the distribution of water content. The pixel number of this picture is 100 x 50. For resolution testing, the imaging object is replaced by a metal plate with different width slot. The testing result is shown in Fig. 8. The testing indicates that the resolution of the proposed electronic THz transmissive system is 2 mm.

IV. CONCLUSION

An electronic transmissive THz imaging system module is implemented by the self-designed integrated circuits and commercial components. The radiated signal of the 40 nm CMOS signal source is modulated by the 100 kHz

chopping signal via commercial SPST switch. The THz signal passes through the object and is received by the 0.18 μm CMOS signal sensor. The received power level is indicated by the sensor and demodulated by the analog multiplier and low-pass filter based chopping signal demodulator. The image resolution of this system is 2 mm.

ACKNOWLEDGEMENT

The authors wish to acknowledge the Ministry of Science and Technology, Taiwan, for funding support, under the grant, NSC-103-2220-E-009-018, Chip Implement Center (CIC) and Industrial Technology Research Institute (ITRI), Hsinchu, Taiwan, for chip implement support and the ANSYS, Taipei, Taiwan, for design support.

REFERENCES

[1] Lei Hou, Hongkyu Park, and Xi-Cheng Zhang, "Terahertz Wave Imaging System Based on Glow Discharge Detector," *IEEE J. Selected Topics in Quantum Electronics*, vol. 17, no.1, pp. 177-182, Jan. /Feb. 2011.

[2] Erik Öjefors, Ullrich R. Pfeiffer, Alvydas Lisauskas, and Hartmut G. Roskos, "A 0.65 THz Focal-Plane Array in a Quarter-Micron CMOS Process Technology," *IEEE J. Solid-State Circuits*, vol. 44, no. 7, pp. 1968–1976, July 2009.

[3] Ruonan Han, et al., "Active Terahertz Imaging Using Schottky Diodes in CMOS: Array and 860-GHz Pixel," *IEEE J. Solid-State Circuits*, vol. 48, no. 10, pp. 2296–2308, Oct. 2013.

[4] Mehmet Uzunkol, Ozan D. Gurbuz, Fatih Golcuk, and Gabriel M. Rebeiz, "A 0.32 THz SiGe 4x4 Imaging Array Using High-Efficiency On-Chip Antennas," *IEEE J. Solid-State Circuits*, vol. 48, no. 9, pp. 2056–2066, Sep. 2013.

[5] Klaus Schmalz, et al., "245 GHz SiGe Transmitter with Integrated Antenna and External PLL," in *Proc. IEEE Int. Microwave Symp. (IMS)*, 2013.

[6] Chun-Hsing Li *et al.*, "A 37.5-mW 8-dBm-EIRP 15.5-degree-HPBW 338-GHz THz Transmitter Using SoP Heterogeneous System Integration," submitted to *IEEE Trans. Microw. Theory Tech.*

[7] SR830 User Manuals, Stanford Research Systems, Inc. [Online]. Available: http://www.thinksrs.com/

[8] Chih-Wei Lai, Wei-Cheng Chen, Tzu-Chao Yan, Chun-Hsing Li, and Chien-Nan Kuo, "The Experimental Study of THz Image Sensor in 0.18 μm CMOS Technology," submitted to *IEEE Asia-Pacific Microwave Conference 2014*.

Intermittently Operating RF Frontend with 5ns Startup Time for 10Gbps Proximity Wireless Communication

Naoki Kitazawa, Kaoru Kohira, and Hiroki Ishikuro

Department of Electronics and Electrical Engineering, Keio University, Yokohama, Kanagawa, Japan
kitazawa@iskr.elec.keio.ac.jp

Abstract — This paper presents intermittently operating RF frontend for proximity wireless communication with wireless power supply. To realize power scalability of the RF frontend, intermittent operation is performed by switching the bias of the amplifier. Quick startup circuit is proposed to accelerate the charging time to a decoupling capacitor. The fabricated test chip in 65nm CMOS occupies 0.075mm². The 3dB bandwidth of amplifier ranges from 1.8GHz to 7.2GHz and current consumption is 8mA. The startup time of the circuit is 5ns.

Index Terms — proximity communication, intermittent operation, quick startup

I. INTRODUCTION

New devices using such as the flexible display attract much attention in recent years. In such applications power should also be transferred without cable for the application of wearable devices (Fig.1(a)). These devices demand high speed and low power communication. For example, Full HD movie demands the data rate up to 3Gb/s and 4K movie demands 7.6Gb/s data rate by the standard of DCI. Furthermore, excellent energy efficiency from low speed to high speed is also important because there is various data rate depending on the display resolution.

TransferJet[1] and RFID are one of the standards for wireless proximity interface. Although the TransferJet's data rate is 375Mb/s and its communication range is about 3cm, the data rate is not enough for display application and it doesn't support wireless power supply.

Interference from power transmission channel to the antenna for communication is a serious problem when the data and power are transmitted simultaneously. Fig. 1(b) shows the measured output signal of amplifier when there is cross-talk from power transmission channel to the data transceiver antenna. Although, low frequency such as 6.78MHz is used for power transmission[2], the communication quality deteriorates if the harmonics are mixed into the band used for communication.

Using OFDM[3] is one of the solution to avoid interference from the power transmission. In the OFDM, the quality of the communication is maintained by eliminating the sub-carrier if there is a strong interference. However, large power is consumed for communication using OFDM. TDMA[4] can be used as another solution to avoid interference. Since there is a periodicity in the

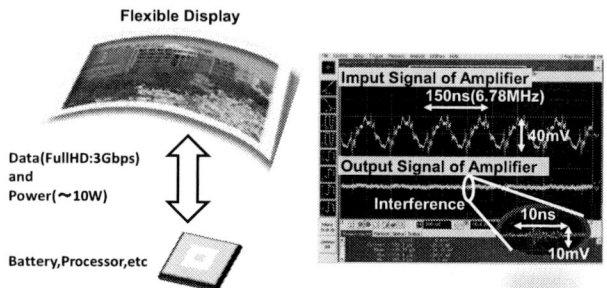

Fig.1 (a)High-speed wireless proximity communication
(b)Output signal of amplifier with interference

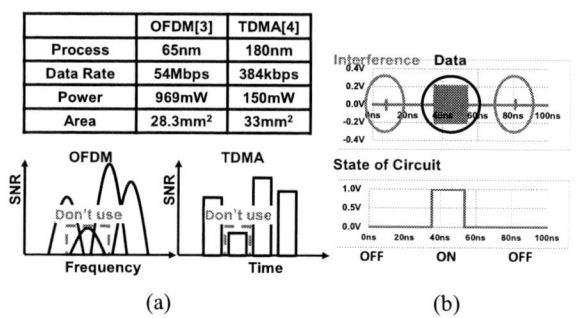

	OFDM[3]	TDMA[4]
Process	65nm	180nm
Data Rate	54Mbps	384kbps
Power	969mW	150mW
Area	28.3mm²	33mm²

(a)   (b)

Fig.2 (a)Comparison of OFDM and TDMA
(b) Intermittent operation

interference from wireless power supply, it is relatively easy to avoid the interference by using TDMA. Fig. 2(a) shows the comparison of the TDMA and OFDM.

In this work, we have developed a proximity wireless interface using near-field with maximum data rate of 10Gbps. To realize the excellent energy efficiency in various data rate and avoid the interference from the wireless power transfer channel, quick startup RF frontend was proposed for intermittent operation.

II. DEMANDS AND ISSUES OF QUICK STARTUP

To realize intermittent operation, bias switching technique had been already reported [5]. However, PSRR is degraded if the size of decoupling capacitor in the bias circuit is not enough. When power is transmitted wirelessly, the power supply can be noisy. Therefore, placing enough

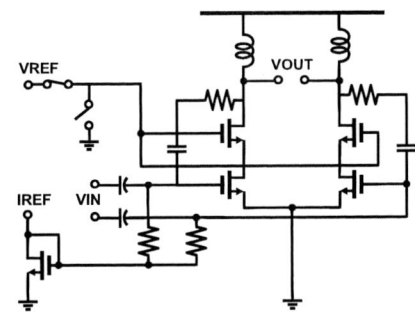

Fig.3 Conventional method for on-off control

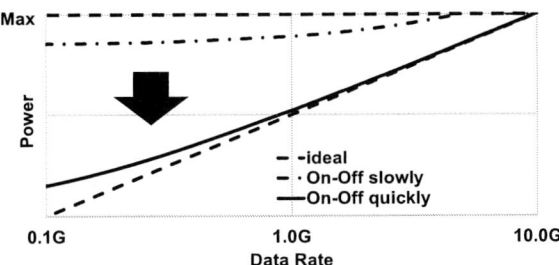

Fig.4 Comparison of power consumption vs. data rate

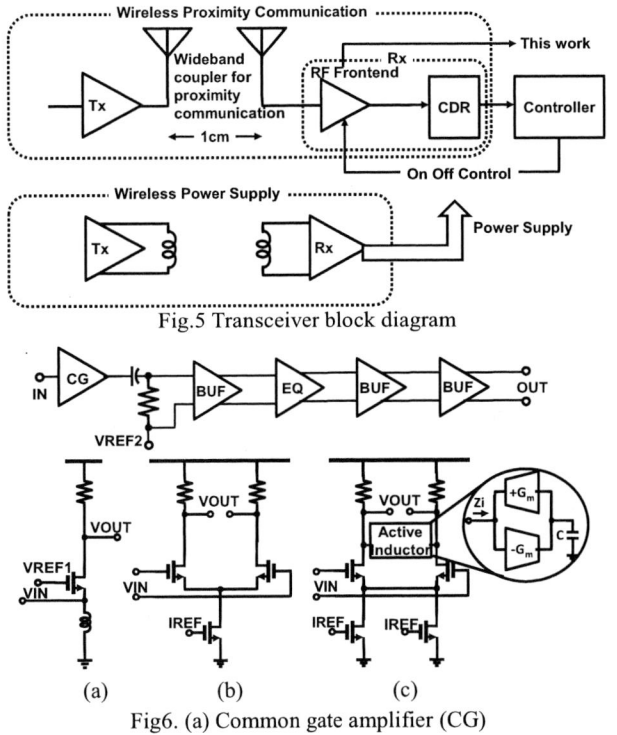

Fig.5 Transceiver block diagram

Fig6. (a) Common gate amplifier (CG)
(b) Buffer stage (BUF)
(c) Peaking stage (EQ)

size of decoupling capacitor in the bias circuit is essential. If the size of the capacitor is large, it takes a long time to charge the capacitor, and therefore, it takes a long time to turning on and off the circuit. It determines the overhead of the packet length of TDMA system and the power consumption is no longer reduced even when the data rate is reduced (Fig.4). In our group, intermittent operation was implemented by turning on and off the bias of current mirror [6]. However, the startup time is too long for 10Gb/s operation.

III. PROPOSED TRANSCEIVER SYSTEM

A. Transceiver Block Diagram

Fig.5 shows the block diagram of the entire transceiver. Data communication is carried out by using wideband coupler for proximity communication. The baseband communication is adopted because simple and low power transceiver can be used. Received signal is amplified by a proposed wideband amplifier and is recovered by the clock and data recovery (CDR) circuit. Power transmission can be carried out using 6.78MHz carrier by inductive coupling technique [2].

B. Proposed Quick Startup Amplifier

Fig.6 shows the schematic of RF frontend amplifier. Wideband input impedance matching is realized by using a common gate (CG) amplifier (Fig6.(a)). Output signal of the CG stage is converted to differential signal by buffer

stage (Fig.6(b)). A peaking circuit with active inductor (Fig.6(c)) extends the amplifier bandwidth. This circuit also attenuates low frequency signal and suppress the interference from wireless power transfer channel. Whole RF frontend amplifier circuits can be quickly turned on and off by switching bias using quick startup circuit. Therefore, the power consumption can be reduced by intermittent operation.

Fig.7 shows schematic of quick startup circuit. This circuit consists of a bias circuit and a comparator. To turn off the amplifier, the gate of transistor (node B) is connected to ground. To turn on the amplifier, node B is connected to current mirror of the bias circuit (node A).

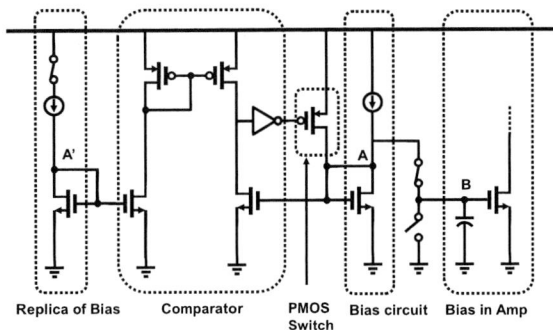

Fig.7 Quick startup circuit

Since the voltage at node A decreases, a comparator compares it with replica of bias voltage (node A'). The PMOS switch is turned on when voltage of node A is lower than that of node A'. Then, decoupling capacitor on node B is directly charged from power supply. When charging is completed, the voltage of node A' and node B are equal. Then, the PMOS switch turns off since the output of comparator turns high.

Without care, this quick startup circuit begins to oscillate. In this design, the parameter is carefully determined to avoid the oscillation even there is a PVT variation.

IV. MEASUREMENT RESULTS

Fig. 8 shows a photograph of the fabricated chip. The area of core chip is $0.075mm^2$. Fig. 9 shows S parameter and NF of the whole RF frontend amplifier. The gain of the amplifier is 22.5dB and the 3dB bandwidth is 1.8 GHz to 7.2 GHz. The measured bandwidth is 15% lower than the designed 3dB bandwidth from 3.5GHz to 8.5GHz. Without intermittent operation, the average current consumption is 8mA at power supply of 1.0V. Fig 10(a) shows comparison of output of amplifier between without and with proposed circuit. Whereas the amplifier takes 30ns to startup without proposed circuit, it is accelerated to 5ns to startup with proposed circuit. The startup time of the circuit is improved 80%. When RF frontend amplifier works intermittently, the power consumption is calculated shown as Fig.10(b). The overhead of power consumption is 4% of max power consumption with proposed circuit while it is 21% without proposed circuit.

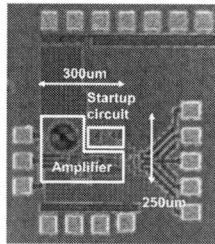

Fig.8 Chip photo

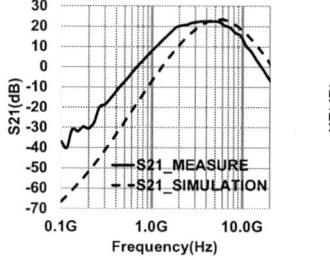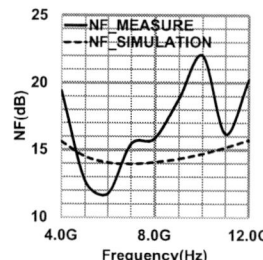

Fig.9 S21 and NF of the amplifier

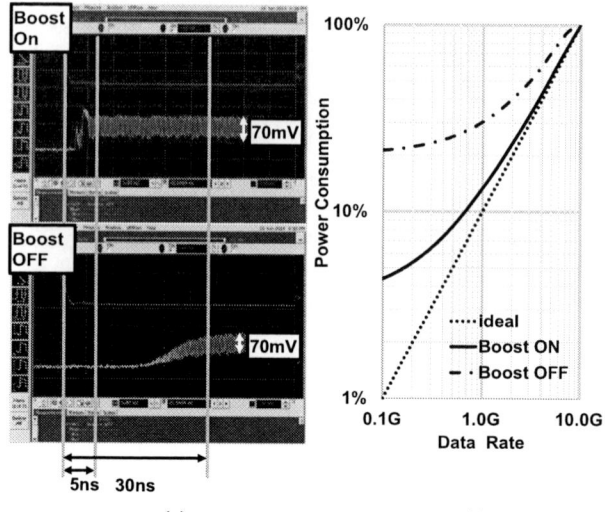

(a)   (b)

Fig.10 (a)Comparison of output of amplifier between w/o and w/ proposed circuit
(b) Comparison of power consumption vs. data rate

V. CONCLUSION

Quick startup circuit of intermittently operating RF frontend for proximity wireless communication has been developed in 65nm CMOS. By using this circuit, the startup time of amplifier with decoupling capacitor is improved by 80%. The overhead of power consumption is improved from 21% to 4% of max power consumption.

REFERENCES

[1] http://www.transferjet.org/, Transferjet, sony, 2014
[2] K. Tomita, H. Ishikuro, et al., "1W 3.3V-to-16.3V Boosting Wireless Power Transfer Circuits with Vector Summing Power Controller," *IEEE Journal of Solid-State Circuits*, pp.2576-2585, Nov.2012.
[3] Chia-Jun Chang, et al., "A CMOS Transceiver with internal PA and Digital Pre-distortion For WLAN 802.11a/b/g/n Applications," *Radio Frequency Integrated Circuits Symposium*, 2010
[4] W.W.Si, et al., "A 1.9-GHz Single-Chip CMOS PHS Cellphone," *IEEE Journal of Solid-State Circuits*, pp. 2737 – 2745, Dec. 2006
[5] T.Terada, et al., "Intermittent Operation Control Scheme for Reducing Power Consumption of UWB-IR Receiver," *IEEE Journal of Solid-State Circuits*, pp. 2702 – 2710, Oct. 2009
[6] Teruo Jyo, Tadahiro Kuroda, and Hiroki Ishikuro, "A 0.8V 1.1pJ/bit inductive-coupling receiver with pulse extracting clock recovery circuit and intermittently operating LNA," *Radio and Wireless Symposium*, pp. 217 – 219, Jan. 2013

978-1-4799-8198-4/15 $31.00 © 2015 IEEE

RFSOI Programmable Array of Capacitors

M. Granger-Jones[1], J. Bendixen[3] J. Costa[2], M. Carroll[2], D. Kerr[2], C. Iversen[3], P. Mason[2], and E. Spears[2]

[1]Qorvo, San Jose, CA, 95134, USA
[2]Qorvo, Greensboro, NC, 27409, USA
[3]Qorvo, Denmark Design Center, Aalborg

Abstract — Tunable capacitors are increasingly important components in today's state of the art RF front ends. They enable a degree of tuning and configurability that is essential in today's multiband multimode handsets. SOI technology is the technology of choice due to the combination of the high degree of integration, high performance and low cost. This paper highlights some of the techniques and tradeoffs involved in the design of a programmable array of capacitors (PAC). The performance metrics are highlighted and the implication of over specification on obtainable Q discussed.

Index Terms — RF CMOS, SOI, RF power switches, tunable networks, tunable capacitor arrays, DTC, PAC.

I. INTRODUCTION

RFSOI solutions on high-resistivity-silicon substrates provide excellent RF performance, ease of design and implementation, extremely high process yields and lower intrinsic cost and a high level of integration not available with other competing technologies. The level of integration afforded by the all silicon implementation allows for relatively complex RF subsystems to be created that combine high linearity RF switches with programmable arrays of capacitors PACs to provide ever increasing degree of re-configurability and tunability required in today's 4G smartphone's RF front ends.

The development history for RFSOI on high-resistivity-silicon technologies has been discussed in detail in several references in the literature [1]. The key innovation that allowed RFSOI switches and tunable products to penetrate the challenging specifications of 4G RF front end applications was the understanding and mitigation of substrate-induced harmonics and distortion products. Today's high performance RFSOI technologies offered by several silicon foundries employ either a combination of harmonic suppression treatments [2] or substrates which contain a 'trap-rich' interface [3]; either one of these approaches has yielded NFET devices capable of being stacked together yielding highly linear switch branches, which is the key building block for all of the RF power and tunable capacitor solutions.

II. PROGRAMMABLE CAPACITOR ARRAYS

A. The Basic SOI CMOS Switch and PAC design

The basic building block for all RF tunable is the RFSOI switch branch. The switch branch in a silicon RFSOI technology is built by cascading in series a number NFET

transistors to create the stacked structure shown in the figure1. Today, nearly all deployed RFCMOS-on-SOI switch solutions use a 0.25um NFET 'output' device built on either a 0.18um or a 0.13um SOI CMOS technology.

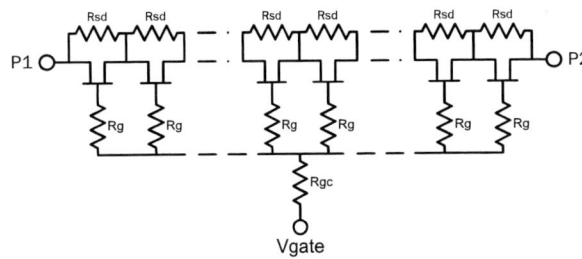

Figure 1) Typical SOI stacked FET configuration for an RF switch

The most common type of programmable array of capacitors (PAC) is a parallel array of binary weighted capacitors in series with high dynamic range RF switches, Figure 2.

A series array of capacitors can also be constructed. This has quite different Q and dynamic range characteristics verses tuning when compared to the parallel array. These properties may be advantageous in certain applications. For example, the parallel array's signal handling capability is largely independent of tuning and lends itself to use in parallel resonant circuits. The series array's signal handling is inversely proportional to capacitance and therefore lends itself to the tuning of series resonances.

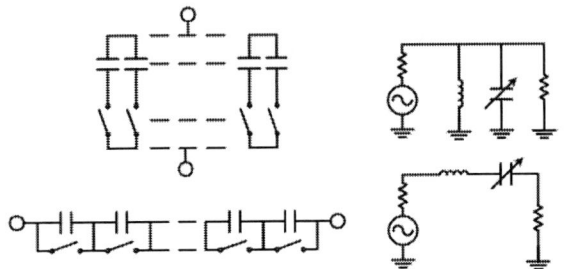

Figure 2 Parallel and Series tunable capacitor arrays and applications.

The RF switches in these PACs would typically be biased in a similar manner to stand-alone RF switches, using gate switching between +VDD/VNEG in the ON/OFF states respectively. The drain and source voltage are biased at 0V. If body contacted FETs are used (not shown) then the body would be biased through high value resistors and typically be switched between 0V/VNEG in the ON/OFF states respectively.

B. PAC Architecture without Negative Voltage Biasing

A PAC biasing scheme is that uses gate/drain/source switching and dispenses with the VEG negative voltage biasing requirement is shown in figure 3 [4]. The bias voltages for the FETs' gates and drain/source are respectively +VDD/0V in the ON state and 0V/VDD in the OFF state. In the case where body contacted FETs are used they would be biased at 0V through high value resistors.

The voltages across the FET junctions are identical to the standard switch biasing approach using a negative voltage VNEG equals VDD. Therefore the two PAC architectures have very similar obtainable Qs and dynamic ranges. This biasing technique is particularly advantageous if charge pump spurs from the VNEG generator are of concern or if the designer does not wish to invest in the additional design overhead of the VNEG charge pump circuitry.

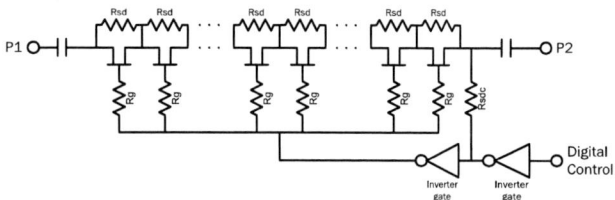

Figure 3 PAC biasing scheme requiring no negative bias voltages

C. PAC Q

The PAC Q is a function of tuning range C_{max}/C_{min} and the $R_{on}*C_{off}$ FOM of the RF switch process used (1) [5]. By restricting the PAC's tuning range, Qs of greater than 100 are obtainable using the most advanced RFSOI processes.

$$Q_{min} = \frac{1}{\omega \left({C_{max}}/{C_{min}} - 1 \right).R_{on}.C_{off}} \qquad (1)$$

The Q_{min} equation assumes that the C_{min} is solely determined by the C_{off} of the switches and that the FETs have been sized to meet the target C_{min}. In moderate Q applications the target C_{min} can be achieved by placing a high quality MIM or MOM capacitor in parallel to a PAC structure. This gives only a modest Q improvement compared to scaling the FET switch size but has the benefits that the C_{min} correlates closer to C_{max} and the die area is not dominated by a huge switch.

D. PAC Dynamic Range

A PAC's dynamic range and signal handling is largely determined by the degree of stacking used in the switch design, the FOM of the process used, the non-linearity of the capacitors and eventually substrate distortion.

In a properly design switch doubling the degree of stacking in conjunction with doubling the width of the FET switches will result in a 6dB increase in signal handling without degrading the PAC Q. If the degree of stacking

exceeds around a factor of 10 this property eventually no longer hold.

The 2nd order capacitor non-linearity can be cancelled by using anti-parallel capacitors. The 3rd order capacitor distortion can be improved by using thicker dielectric or by applying the stacking technique to the capacitor structures.

III. PAC APPLICATIONS

E. Stand Alone PACs

Stand alone RFSOI PAC are a useful versatile component that find applications in many tuning applications. A 4 bit binary weighted PAC design that provides a 0.5pF-5pF tuning range is shown in Figure 4. It contains all of the digital/analog circuitry, including the serial interface (GPIO or RFFE), charge pump, ESD network and switch drivers to provide extremely small and cost effective tuning solutions. This very compact and highly integrated design and other similar configurations are being shipped in extremely high volume in antenna applications in the cellular handset market.

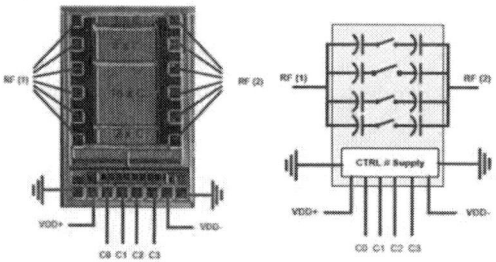

Figure 4) Tunable array of capacitor built with RFSOI switch branches and MIM capacitors

The measured performance for a 4bit PAC on using an SOI process with a 220fs FOM is shown in figure 5. The device's Q falls short of the Q_{min} predicted by (1). This deviation from the ideal is caused by a combination of the loading of the biasing network, routing resistance and parasitic capacitances that reduce the net C_{max}/C_{min} ratio.

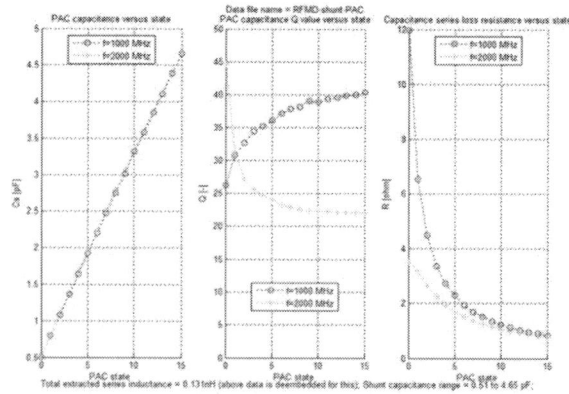

Figure 5) 4bit PAC Capacitance, Q and series Resistance verses tuning state.

978-1-4799-8198-4/15 $31.00 © 2015 IEEE

F. RFSOI Impedance Tuners

The combination of RF switch branches and RF switchable MIM capacitors can be extended even further to realize high-Q impedance tuning networks to optimize the power delivered to a cellular handset antenna under a myriad of different environmental conditions. Typical impedance tuner solutions include a combination of external high Q inductors working in conjunction with the internal tunable MIM capacitors and a combination of RF switches to efficiently synthesize a number of different impedance states which optimize the impedance match of a cellular antenna to the other components of the RF front end.

A great advantage of impedance tuners built with RFSOI technology is that the integration of low loss bypass RF switch networks becomes trivial, something that is not easily accomplished with tunable capacitor MEMS arrays solutions available commercially today. The figure 6 depicts a broadband SOI impedance tuner shipping in very high volume in several leading 4G handsets.

Die photo	Application board photo

Figure 6) antenna tuner implemented with a combination of RFSOI die and external high Q inductors.

The impedance tuner depicted above performs very well in matching and optimizing the varying impedance characteristics to the RF front end module for a given antenna. The net gain or loss of this particular RFSOI antenna tuner at 895MHz is illustrated in figures 7. The RF antenna tuner gain charts shows an insertion loss of 0.5dB at 1:1 VSWR. This loss however is recovered very quickly with VSWR's in the >=2 range, and as VSWR increases, the tuner realizes net positive gain for the overall network showing a clear improvement in typical antenna implementations.

IV. CONCLUSION

RFSOI on high-resistivity-silicon solutions for the cellular handset industry are being deployed in extremely large volumes by the leading RF front end providers.

The combination of highly linear RF switches are combined with tunable capacitor arrays to provide the reconfigurable and tunable designs required to accommodate the complexity of the RF front end in leading 4G handsets.

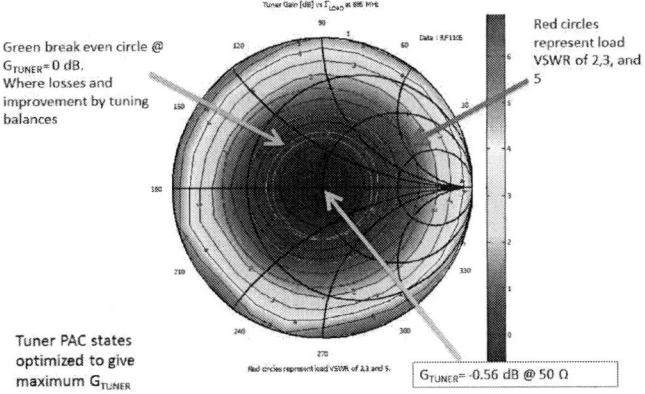

Figure 7) Low Band (895MHz) impedance tuner performance. The colors in the circle represent the effective gain/loss of the combined impedance tuner and antenna. At VSWR's levels approaching 2, the tuner starts providing a net performance gain for the overall circuit and at higher VSWR levels, the gain improvement becomes very significant.

REFERENCES

[1] Costa,et al, "RFCMOS SOI Technology for 4G reconfigurable RF solutions" , *International Microwave Symposium, June 2013.*

[2] Botula, A. et al, "A Thin-Film SOI 180nm CMOS RF Switch Technology," *Silicon Monolithic Integrated Circuits in RF Systems, 2009. SiRF 2009.* pp.1-4, 19-21 Jan. 2009

[3] Kerr, D.C et al. , "Identification of RF Harmonic Distortion on Si Substrates and its Reduction Using a Trap-Rich Layer," *SiRF 2008.* pp.151-154, 23-25 Jan. 2008

[4] Donggu Im ;   Kwyro Lee "Highly Linear Silicon-on-Insulator CMOS Digitally Programmable Capacitor Array for Tunable Antenna Matching Circuits" Microwave and Wireless Components Letters, IEEE Volume: 23 , Issue: 12

[5] Whatley, R.; Ranta,T.; Kelly, D. "RF Front-End Tunability for LTE Handset" , *CSICS, 2010 IEEE*

Improvements in SOI Technology for RF Switches

Mark Jaffe[1], Michel Abou-Khalil, Alan Botula, John Ellis-Monaghan, Jeffrey Gambino, Jeff Gross, Zhong-Xiang He, Alvin Joseph, Richard Phelps, Steven Shank, James Slinkman, and Randy Wolf

IBM Microelectronics Division, Essex-Junction, Vermont, 05452, USA.
[1] mjaffe@us.ibm.com

Abstract—**Over the past few years, CMOS Silicon-on-insulator (SOI) has emerged as the dominant technology for RF switches in RF front end modules for cell phones and WiFi. RF SOI technologies were created from silicon processes originally used for high speed logic applications, but the technology was modified to meet the performance needs of RF switches. The RF SOI technologies have been improved to follow the evolving system requirements for insertion loss, isolation, voltage tolerance, linearity, integration and cost. In this paper, the performance results of the latest generations of RF SOI switch technologies from IBM are reviewed and technology elements that contribute to improved performance are discussed. Future improvements are also discussed.**

Index terms—**RF switches, SOI, Front End Module**

I. INTRODUCTION

The RF capability of cell phones has steadily and significantly increased from the very beginnings of mobile phone communications in 1973 when Martin Cooper placed the first mobile phone call, through the 1980's with 1G cell service and 10kbs data rates to current LTE smartphones with 100mbs data rates. This evolution in cell phone RF capability has been mirrored by improvements in cost and capability of the individual components. The RF switch found in the front end module's switch has evolved from GaAs PHEMT switches [1] to a more integrated solution with Silicon CMOS on sapphire [2] to a lower cost and high performing Silicon on Insulator solution [3]. The evolution in phone standards has increased the number of RF switch ports from about 6 for the 2G standard to around 30 for today's 4G LTE handsets [4]. This necessitates improvements in linearity, insertion loss and isolation to enable this evolution.

In this paper, we report on improvements in the SOI technology used to fabricate RF switches. We present a brief description of device technology and show a summary of both DC and RF measurement data.

II. DEVICE TECHNOLOGY

IBM qualified its first RFSOI technology, CSOI7RF, in 2008 by adapting a 180nm high speed logic technology which had been used to make microprocessors. The p-type handle substrate was changed from 10 Ω-cm to a high resistivity (~1k- Ω-cm) and the buried oxide was thickened to 1um to reduce substrate parasitic coupling. The device silicon thickness was maintained at 0.145um [5]. The NFET was optimized for RF switch applications. This technology, offers a figure of merit Ron*Coff of 210fs and a switch biased breakdown voltage of 3.8V. CSOI7RF has enjoyed tremendous market success and is currently the highest volume switch technology in the marketplace.

IBM improved on this technology over time by decreasing the channel length of the switch transistor. The gate length of a logic NFET is limited by conducting hot electron reliability. A minimum gate length of 0.32um is tyipcal for 2.5V transistors. The switch transistor is never biased in saturation – it is either on in the linear mode or off. This allows a shorter channel length to be used as only non-conducting hot electron reliability needs to be considered. This allowed IBM to qualify a 0.28um and a 0.24um switch FET. Reducing the channel length lowers Ron without changing Coff. IBM was able to achieve a Ron*Coff of 165fs for the 0.24um switch NFET. Further reductions in the channel length were limited by decreasing breakdown voltage.

In 2013, IBM reengineered the switch NFET to further improve switch performance metrics. Device design for switches is a difficult tradeoff as virtually everything which reduces Ron also tends to reduce breakdown voltage and increase Coff. By both thinning the transistor level silicon and redesigning the junctions, IBM was able to maintain Ron and decrease Coff as well as increase breakdown voltage. The new transistor was implemented by a patented technique to achieve localized body thinning within the switch NFET active area, where body thickness is thinned down to 100nm. Within this technique, the new transistor can be combined with the existing logic and circuit designs of the base technology. This transistor, which was called the LOWD switch FET, demonstrated a Ron*Coff of 140fs and a breakdown voltage at a channel length of L=0.24um of 4.7V [5].

In 2014, IBM created a new technology called CSOI7SW. This is a hybrid technology using features from both the 0.18um and the 0.13um lithography

978-1-4799-8198-4/15 $31.00 © 2015 IEEE

generations. The hybrid approach allows the density of a 0.13 technbology in the switch region of the chip which is usually the majority of area of a chip while maintaining the thicker metals of 0.18 technologies for power handling and lower resistance. This technology has an ultra-thin body of 80 nm to help reduce Coff (< 215 fF/mm) and uses a thicker gate oxide than prior IBM switch technologies to help reduce Igidl and leakage, while increasing breakdown voltage and supporting optional higher gate bias for improved performance. The switch transistors in this new technology exhibit Ron*Coff from 128fS down to 115fS depending on biasing and layout options.

Table I shows the DC breakdown voltage for the switch devices, comparing all three technologies, for 3 different gate lengths. BVdss is measured for Vg=Vb=0, whereas BVoff is measured at Vg=Vb=-2.5V. While all device designs exhibit roll off in breakdown voltage with decreasing channel length, the LOWD and CSOI7SW devices were engineered to have the same BVoff at the shortest channel length, 0.24um, as CSOI7RF has at 0.32um.

TABLE I: BVDSS AND BVOFF FOR
DIFFERENT TECHNOLOGIES AND DIFFERENT LG

	Device	0.24um	0.28um	0.32um
BVdss (Vg=Vb=0V)	Base	2.75	3.4	3.8
	LOWD	2.9	3.6	4.2
	SW	3.8	3.9	4.6
BVoff (Vg=Vb=-2.5V)	Base	3.7	4.3	4.7
	LOWD	4.7	5.2	5.5
	SW	4.7	5.1	5.5

III. RF DATA

RF test structures were designed to measure both s-parameters and harmonics. Series and shunt elements with 6 and 12 stack were put into RF pad sets. Series circuits were used to extract Ron and Coff as well as on state harmonics. Shunt elements were used to measure off state harmonics under near matched conditions. Deimbedding structures allowed the removal of the pads and top level metal leaving effects of the first two levels of metal as these are always a part of the switch layout.

Figure 1 shows Ron*Coff dependency on the gate length of the switch device. The LOWD technology significantly reduced Coff from the CSOI7RF technology. Reducing the channel length reduces Ron. The CSOI7SW technology with its higher allowed gate-on

bias reduced Ron even further. Figure 2 shows the dependency of insertion loss on the operation frequency.

Linearity data is shown in figures 3 and 4. These measurements were taken for 12stack switch structure with 1mm total width. The figures show comparisons of the 2nd and 3rd harmonics for LOWD, CSOI7SW, and CSOI7SW with higher gate biasing for the series and shunt structures. Switch harmonics are created by nonlinearities in the transistor and parasitic loading from the substrate. IBM uses a proprietary process to reduce substrate loading which is called the "BTQ" process (i.e. a crystal damage implant into the substrate through a trench) [5] while most of the rest of the industry uses trap rich substrates. The samples in figures 3 and 4 had the BTQ process.

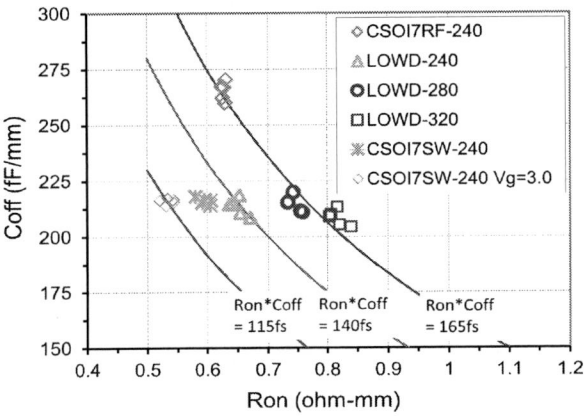

Fig-1: Ron*Coff for a 1mm switch of different IBM technologies

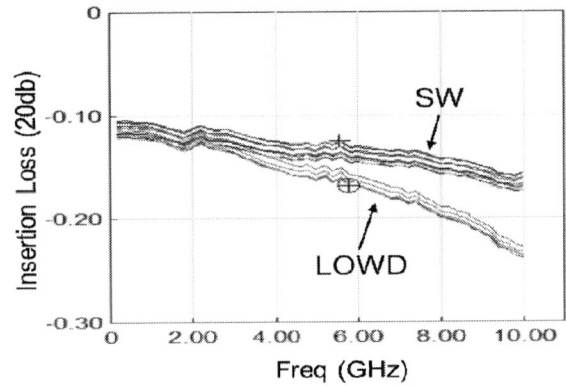

Fig-2: Insertion Loss vs frequency for the LOWD and the CSOI7SW technologies

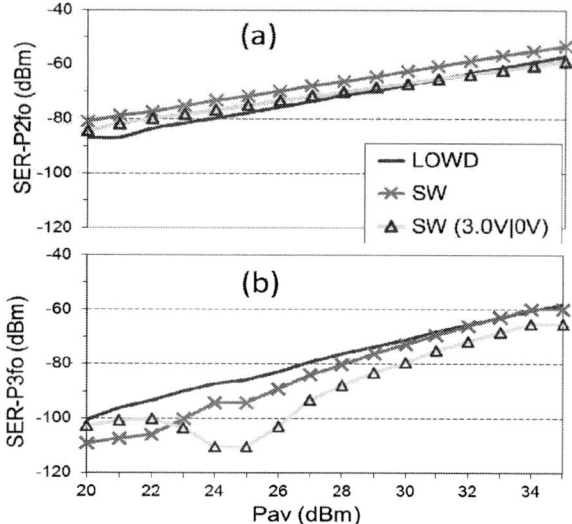

Fig-3: (a) 2nd harmonic and (b) 3rd harmonic data vs input power for 12 stack 4mm series switch leg structure.

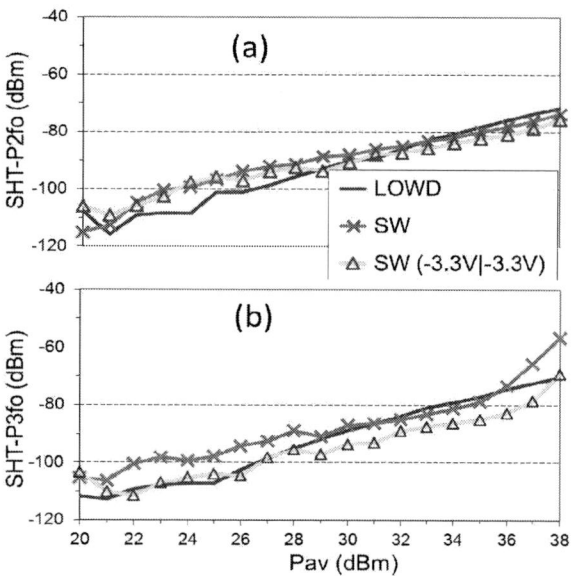

Fig-4: (a) 2nd harmonic and (b) 3rd harmonic data vs input power for 12 stack 4mm shunt switch leg structure.

IV. FUTURE TRENDS

The requirements on components in the front end module continue to increase. Architectures in which the signal traverses multiple switches between the PA and the antenna drive the need for lower insertion loss. Carrier aggregation requires improved linearity. SOI Switch technology will continue to improve to meet these needs. OI device design will continue to optimize the tradeoffs between Ron and breakdown voltage and to reduce the total device non-linearities. Improved substrate technologies will also help to reduce substrate loading which will reduce both harmonics and insertion loss.

V. CONCLUSION

We reported of improvements in the SOI technology used to fabricate RF switches. We presented a brief description of device technology and show a summary of Break down voltages, RonCoff obtained from s-parameter tests, and harmonics measurements obtained switch element test structures. The newly developed technology, CSOI7SW, shows a significant improvement in Ron and frequency dependency, while maintaining low Coff, breakdown voltage and harmonic distortion as the previous IBM RF SOI technology.

ACKNOWLEDGEMENT

The authors would like to thank the in line test team, and the DC and RF test teams for their helpful assistance to generate the data needed for this paper.

REFERENCES

[1] Lucero, R et al., "Design of an LTCC switch duplexer front-end module for GSM/DCS/PCS applications," RFIC Symposium, 2001.
[2] Bonkowski, J et al., "Integration of triple-band GSM antenna switch module using SOI CMOS," RFIC Symposium, 2004.
[3] R Wolf et.al. "A Thin-film SOI 180nm CMOS RF switch", SiRF 2009
[4] J.E.Mueller et.al, " Requirements for reconfigurable 4G Front-Ends," IEEE International Microwave Symposium, 2013
[5] A. Botula et.al, "A thin film SOI 180 nm CMOS RF Switch Technology," SiRF 2009
[6] A. Tomba, "Silicon-on-Insulator Switches for Cellular and WLAN Front-End Applications", RFIC 2012

High Resistivity SOI wafer for mainstream RF System-on-Chip

Jean-Pierre Raskin[1] and Eric Desbonnets[2]

[1] Université catholique de Louvain (UCL), Institute of Information and Communication Technologies, Electronics and Applied Mathematics, Place du Levant, 3, B-1348 Louvain-la-Neuve, Belgium, Phone: +32.10.47.23.09, jean-pierre.raskin@uclouvain.be

[2] Soitec, Parc Technologique des Fontaines, 38190 Bernin, France, eric.desbonnets@soitec.com

Abstract — The increasing demand for wireless data bandwidth and the rapid adoption of LTE and LTE Advanced standards push radio-frequency (RF) IC designers to develop devices with higher levels of integrated RF functions, meeting more and more stringent specification levels. The substrates on which those devices are manufactured play a major role in achieving that level of performance [1]. In this paper, UCL and Soitec explain the value of using RF-SOI substrates and more especially the new generation of Soitec widely adopted eSI™ (enhanced Signal Integrity) substrate to achieve the RF IC performance requested to address the LTE Advanced smart phone market.

Index Terms — Silicon-on-Insulator (SOI), high resistivity Si substrate, trap-rich layer, high frequency, wireless applications, LTE.

I. INTRODUCTION

Growing usage of multimedia applications associated with consumers' desire for ultimate mobility experience has created significant market changes:

i.   More mobile devices: The number of smart phones will grow at a compound annual growth rate (CAGR) of 12.7 percent between 2013 and 2018 (Source: IDC [2]).

ii. More mobile users: The number of mobile subscriptions worldwide exceeds world population since 2014 and will reach 9.2 billion by 2019. More than 65% of the world population will be covered by LTE in 2019 versus 25% today (Source: Ericsson [3]).

iii. More multimedia applications: Driven by video streaming, TV, social media and content sharing, etc., data traffic continues to grow at a 61%CAGR (Fig. 1 [4]).

To respond to this data traffic demand, faster mobile throughput is required. Carrier Aggregation (CA) of bands and multi input multi output (MIMO) techniques are the main hardware techniques in handset to increase capacity. For example, in LTE Release 8, theoretical maximum data rate with 20 MHz bandwidth is 150 Mbs download / 50 Mbs upload. In LTE Release 12, theoretical maximum data rate with 3x20 MHz bandwidth is 1.2 Gbs download with 4x4 MIMO and 0.6 Gbs upload with 2x2 MIMO.

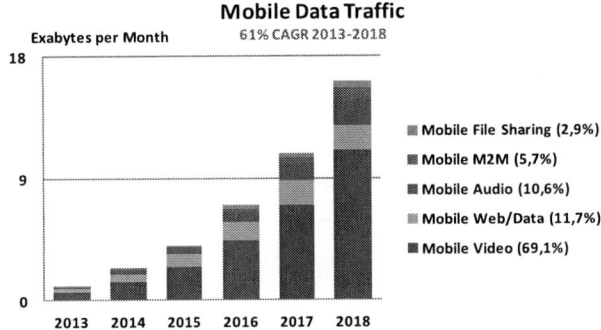

Fig. 1. Cisco VNI Mobile, February 2014 [4].

Common flagship devices are today using carrier aggregation in receive mode with 4 receivers (MIMO received on) and one transmitter operating simultaneously. Most stringent linearity requirements are then required on most critical Front End Module blocks such as main antenna switches, swapping switches, antenna tuners. IIP3 requirements up to 90 dBm are foreseen to handle some specific cases (Table I). For example, carrier aggregation of Band 17 & 4 when Band 17 Tx third harmonic fall into Band 4 Receive.

To address high volume market at competitive cost and meet linearity requirements demand, since 1992 Soitec and UCL has been cooperating to enable a standard CMOS compatible silicon wafer with nearly perfect RF performance: RF Silicon-on-Insulator (RF-SOI).

TABLE I

INTEL MOBILE, "CHALLENGES FOR RADIOS DUE TO CARRIER AGGREGATION REQUIREMENTS", LARRY SCHUMACHER, NOV. 2012 [5].

Network	Linearity (IIP3 in dBm)
2G	55
3G	65
4G LTE	72
4G LTE + CA	Up to 90

II. RF-SOI SUBSTRATE – BEST PERFORMANCE-COST TRADE-OFF

RF-SOI foundries are constantly improving devices performance. A typical figure of merit used for switches is Ron x Coff. We have seen 20% Ron x Coff performance improvement per year for the last 8 years.

The main challenges for RF SOI substrate is to reduce its contribution to the non-linear behaviour of the RF IC compared with an ideal substrate (fully dielectric). When measuring the same RF IC onto different substrates from ideal to very nonlinear substrates, substrate target specifications must fall into the "IC limited" zone (Fig. 2).

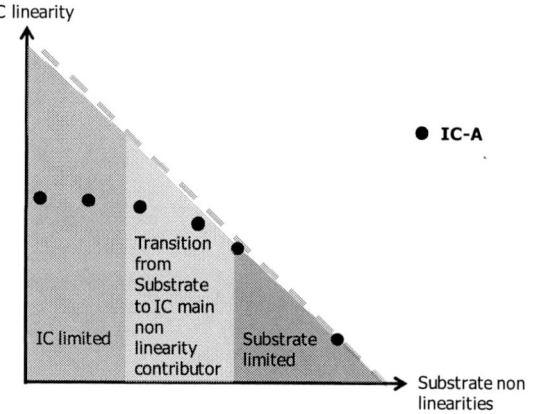

Fig. 2. Circuit performance versus substrate non-linearity.

Since 2008, by offering the best cost, area and performance to the market, RF-SOI has progressively displaced GaAs and Silicon-on-Sapphire technologies has become the mainstream technology for antenna switch with more than 85% market share (Yole, February 2014, [6]) and is available in all foundries.

RF-SOI wafers share the same challenges as the circuits and devices to meet IIP3 requirements: never be the limiting performance factor.

Beyond performance, RF-SOI offers a unique advantage to further reduce board area by integrating all front end module devices on the same die.

III. eSI SUBSTRATE PERFORMANCE

In 1997, UCL presented pioneering work on the RF performance of high-resistivity (HR) SOI substrate material. The great interest of HR SOI substrate to reduce RF losses as well as crosstalk in Si-based substrates was demonstrated [7]. It has been demonstrated in [8] that HR-Si must present an effective resistivity as high as 3 kΩ.cm to be considered low loss for RF applications. However, oxidized HR-Si substrate suffers from parasitic surface conduction (PSC) due to fixed oxide charges which attract free carriers near the Si/SiO_2 interface, thereby reducing the substrate effective resistivity (ρ_{eff}) of the wafer by more than one order of magnitude compared with the bulk nominal resistivity.

In 2005, UCL presented the possibility of creating HR SOI substrates characterized with an effective resistivity as high as 10 kΩ.cm (Fig. 3) thanks to the silicon surface modification below the buried oxide (BOX) of a high resistivity SOI substrate. The surface modification consists in the introduction of a high density of defects called traps at the BOX / HR-Si handle substrate [9]. Those traps originate from the grain boundaries in a thin (300 nm-thick) polysilicon layer. This high-resistivity characteristic, which is conserved after a full CMOS process, translates to very low RF insertion loss (< 0.15 dB/mm at 1 GHz) along coplanar waveguide (CPW) transmission lines and purely capacitive crosstalk (Fig. 4) similarly to quartz substrate.

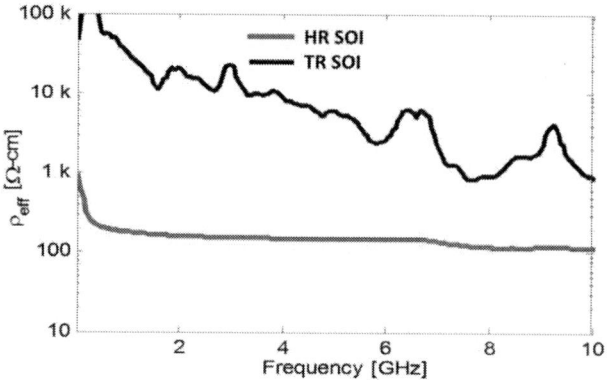

Fig. 3. Measured effective resistivity of a high-resistivity SOI substrate (HR SOI) and a trap-rich (TR SOI, i.e. eSI TR-SOI from Soitec) HR SOI substrate presenting both of them a handle Si substrate characterized by a nominal resistivity of 10 kΩ.cm. While TR SOI presents an effective resistivity of around 10 kΩ.cm, the value of the HR SOI without traps is actually only of 200 Ω.cm.

It has been demonstrated that the presence of a trapping layer does not alter the DC or RF behavior of SOI MOS transistors [10]. Besides the insertion loss issue along interconnection lines, the generation of harmonics in the Si-based substrates has been investigated [11]. UCL demonstrated that harmonics level originated from the substrate is reduced by at least 20 dB moving from standard resistivity SOI substrate (~ 10 Ω.cm) to high resistivity SOI (~ 1 kΩ.cm), and more importantly an additional drop of 40 dB is achieved with the innovative trap-rich HR SOI (TR SOI) substrate, as illustrated in Fig. 5. This low harmonic level is comparable with insulating substrate. The improvement of the HR SOI substrate with the introduction of defects brings also clear benefits for the integration of passives, such as the quality factor of spiral inductors [12], tunable MEMS capacitors [12], as well as for the reduction of the substrate noise (crosstalk) between devices integrated on the same chip (Fig. 6).

978-1-4799-8198-4/15 $31.00 © 2015 IEEE

Fig. 7 presents the noise signal amplitude recorded at the drain electrode of the transistor when the injected noise signal frequency varies from 500 kHz up to 50 MHz and a dc voltage of 0 or -10 V is applied on the noise signal pad. A deep depletion is formed under the BOX when a negative dc bias is added to the clock frequency, therefore the coupled noise decreases by 10 dB comparing with its level at 0 V for HR-SOI substrate whereas there is no change for eSI TR-SOI.

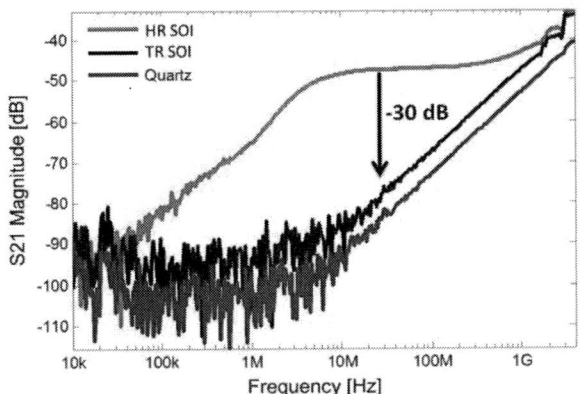

Fig. 4. Measured crosstalk response for the basic test structure on quartz and HR based-Si substrates with (TR SOI, i.e. eSI TR-SOI from Soitec) and without (HR SOI) 300 nm-thick polysilicon trap-rich layer. The distance d between the disturbing input and the output device is 50 µm.

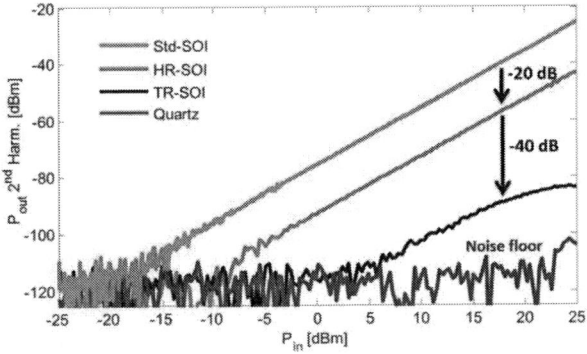

Fig. 5. Measured harmonic distortion along a CPW line lying on standard SOI (~ 10 Ω.cm) and HR based-Si substrates (~ 10 kΩ.cm) with (TR SOI, i.e. eSI TR-SOI from Soitec) and without (HR SOI) trap-rich layer. Identical CPW transmission line has been measured on top of a quartz substrate to quantify the harmonic measurement noise floor.

UCL and Soitec have been working together to identify the technological opportunities to still further improve the high-frequency performance of commercially available HR-SOI substrates. Thanks to the introduction of an engineering substrate handle based on the Prof. Raskin's discovery, Soitec has ramped up in early 2012 a new flavor of HR-SOI called eSI™, for enhanced Signal Integrity substrate with a measured effective resistivity beyond 3 kΩ.cm. Thanks to the introduction of eSI, the RF SOI

substrate can really be considered as a lossless Si-based substrate. Beyond switch, eSI RF-SOI technology opens the path to further system integration in the Front End Module space as well as even more complex mixed signal System-on-Chip (SoC) [13]. In February 2014, Qualcomm has announced the first smart phone having adopted its RF360 solution. Qualcomm unveiled that their "RF360 is based on CMOS and SOI technology". As pointed out by several press releases published this year, the demand for the newly developed HR-SOI is booming and it is becoming the major material for high-performance RF applications. All major foundries in the field of RF applications have adopted the newly developed HR SOI substrate.

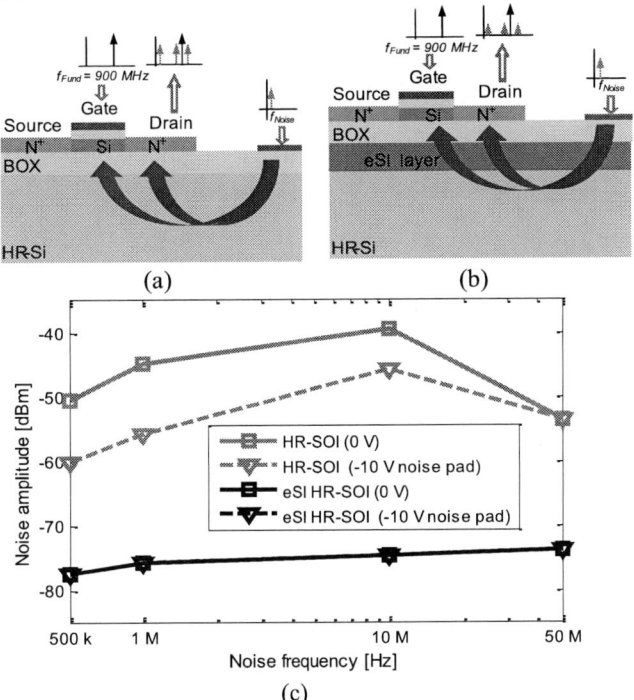

Fig. 6. Cross-section of FD SOI MOSFET with noise pad on a (a) HR-SOI and (b) eSI TR-SOI wafer, (c) Power levels recorded at the transistor drain electrode when biased in saturation and lying on either HR-SOI or eSI TR-SOI wafers when a noise signal is fed into the noise pad and no RF signal is applied to the gate.

The linearity performance of eSI wafers provided by Soitec is presented in Fig. 7 for the first and new substrate generations.

New generation eSI products meet required specifications in terms of losses, coupling and non-linearities to push further the RF ICs performance limit.

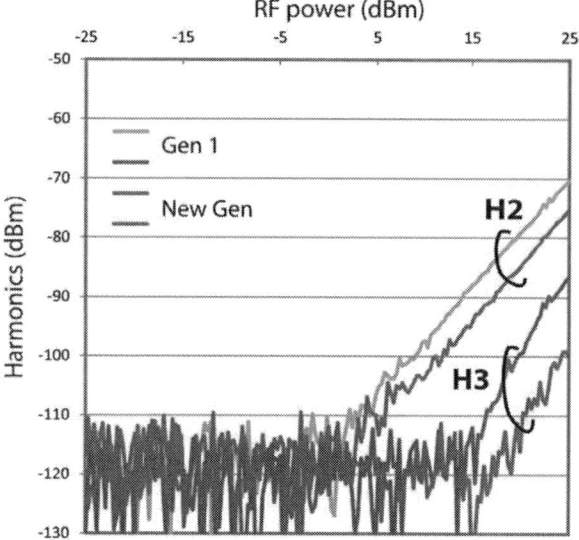

Fig. 7. Linearity performance of commercially available eSi substrates from Soitec.

IV. NEXT CHALLENGES SERVED BY SOI

Pushed by still more data demanding video applications, linearity requirements will continue to be even more stringent, dual carrier upload being the next technological implementation driving current substrate innovations. Beyond its demonstrated market adoption on RF Front End Module, RF-SOI is close to be in 100% of the worldwide smart phone. In parallel, digital Fully Depleted SOI (FD-SOI) is also being adopted by bringing the best cost-performance trade-off at 28 nm and beyond for mobile digital devices such as application processors. It offers up to 10x lower consumption compared with standard bulk CMOS thanks to its capability to operate at much lower supply voltage and lower biasing current for equivalent processing performance. Combining advanced CMOS process with optimized SOI substrate will enable RF and digital integration for advanced SoC integration. SOI is then very well positioned to serve future 5G standards and IoT applications such as wearables that will respectively run on higher bands than today's 4G and will require drastic power consumption reduction.

ACKNOWLEDGEMENT

The authors wish to acknowledge Didier Basso, Cédric Malaquin, Christophe Didier, Bernard Aspar and Communication & Power Product Engineering teams of Soitec for their strong support.

REFERENCES

[1] E. Desbonnets, S. Laurent, "RF Substrate Technologies for Mobile Applications", Soitec White Paper, 2011.
[2] 1.8 billion SmartPhone units in 2018, resulting in a 12.7% compound annual growth rate (CAGR) for the 2013 – 2018 forecast period, IDC Aug. 28, 2014.
[3] Ericsson Mobility Report, June 2014.
[4] Cisco VNI Mobile, February 2014.
[5] Intel Mobile, "Challenges for Radios due to Carrier Aggregation requirements", Larry Schumacher, Nov. 2012.
[6] Sapphire Applications & market: from LED to Consumer Electronic, Yole, 2014.
[7] J.-P. Raskin, A. Viviani, D. Flandre and J.-P. Colinge, "Substrate crosstalk reduction using SOI technology", *IEEE Transactions on Electron Devices*, vol. 44, no. 12, pp. 2252-2261, December 1997.
[8] D. Lederer and J.-P. Raskin, "Effective resistivity of fully-processed SOI substrates," *Solid-State Electronics*, vol. 49, no. 3, pp. 491-496, 2005.
[9] D. Lederer and J.-P. Raskin, "New substrate passivation method dedicated to HR SOI wafer fabrication with increased substrate resistivity," *IEEE Electron Device Letters*, vol. 26, no. 11, pp. 805-807, 2005.
[10] K. Ben Ali, C. Roda Neve, A. Gharsallah, and J.-P. Raskin, "RF performance of SOI CMOS technology on commercial 200 mm high resistivity silicon trap-rich wafers", *IEEE Transactions on Electron Devices*, vol. 61, no. 3, pp. 722-728, March 2014.
[11] C. Roda Neve and J.-P. Raskin, "RF Harmonic Distortion of CPW Lines on HR-Si and Trap-Rich HR-Si Substrates," *IEEE Transactions on Electron Devices*, vol. 59, pp. 924-932, 2012.
[12] Yonghyun Shim, J.-P. Raskin, C. Roda Neve, and M. Rais-Zadeh, "RF MEMS passives on high-resistivity silicon substrates", *IEEE Microwave and Wireless Components Letters*, vol. 23, no. 12, pp. 632-634, December 2013.
[13] E. Desbonnets, C. Didier, N. Pautou, "Innovative RF-SOI wafers for wireless applications", Soitec White Paper, 2013.

978-1-4799-8198-4/15 $31.00 © 2015 IEEE

Comparison of Substrate Effects in Sapphire, Trap-Rich and High Resistivity Silicon Substrates for RF-SOI Applications

Vikram Sekar, Chih-Chieh Cheng, Richard Whatley,
Chang Zeng, Alper Genc[1], Tero Ranta, and Francis Rotella.

Peregrine Semiconductor Corporation, San Diego CA, 92121, USA.

Abstract — This paper compares insertion loss, harmonics and Q-factor of sapphire, trap-rich silicon and high-resistivity silicon substrates. The concept of substrate characteristic frequency is shown to be useful metric to evaluate substrate behavior. New methods to evaluate substrate properties are introduced using open CPW stubs and parallel plate capacitors. It is demonstrated that sapphire provides the best RF performance, while trap-rich silicon offers limited but adequate performance with a resistivity around a few kΩ-cm. High-resistivity silicon provides the worst RF performance due to unmitigated parasitic conduction.

Index Terms — RF silicon-on-insulator, trap-rich silicon, silicon-on-sapphire, resistivity, harmonics, semiconductor.

I. INTRODUCTION

Due to a growing need for high performance CMOS technology in the RF market place, silicon-on-insulator (SOI) devices built on high resistivity substrates have become very attractive for system-on-chip applications. In the 1980s, sapphire was first demonstrated as an excellent substrate for CMOS SOI [1], the result of which has enabled >2 billion RF switches using Silicon-on-Sapphire (SOS) technology. High resistivity silicon substrates (HRSOI) have been considered for RF-SOI, however the parasitic surface conduction (PSC) at the silicon-insulator interface results in high RF losses and poor linearity. To eliminate PSC, a high density of interface traps was introduced to freeze free carriers, resulting in the "trap-rich" HRSOI (TRSOI) technology [2]. Additional techniques involving trenches and blanket implants in high resistivity silicon wafers have been studied to improve effective substrate performance [3].

The goal of this paper is to compare the performance of SOS, HRSOI and TRSOI technologies in terms of loss, linearity and quality-factor. Additionally, this paper presents two new techniques for evaluating substrate properties using: (i) open CPW stubs, and (ii) parallel plate capacitors. This paper also proposes the use of substrate characteristic frequencies as viable metrics to compare the RF performance of SOI substrates. Detailed circuit analysis of substrates, along with EM simulation and measurement data are presented.

---

[1] Now with Entropic Communications, San Diego, CA, 92121.

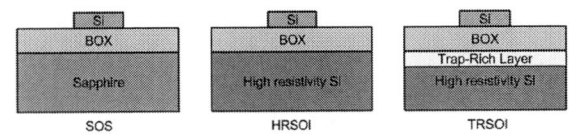

Fig. 1: Cross-sections of silicon-on-insulator technologies.

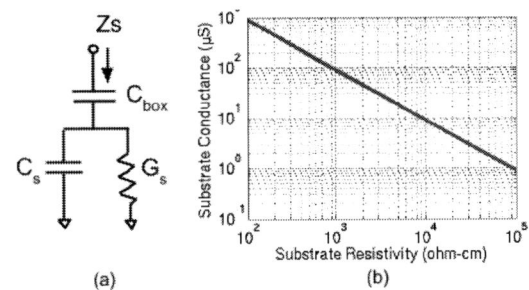

Fig. 2: (a) Simplified circuit model of SOI substrate, and (b) variation of substrate conductance Gs with resistivity of handle wafer.

II. THEORY

Fig. 1 shows the cross-sections of SOS, HRSOI and TRSOI silicon-on-insulator technologies, where there is a layer of buried oxide (BOX) insulating the silicon devices from the handle substrate below. Fig. 2(a) shows the simplified equivalent circuit model of a oxide-semiconductor system [4], where C_{box} is the BOX capacitance per unit area, and C_s and G_s are the equivalent substrate capacitance and conductance respectively, at a given frequency and bias. The dependence of Gs on substrate resistivity is shown in Fig. 2(b) and is extracted using Ansys Q3D extractor assuming a 1mm² plate of metal on oxide, over a 200um thick handle wafer with a ground plane underneath. The simulated capacitance Cs is 520 fF. C_{box} is assumed to be 50pF and depends on the thickness of the BOX layer. Based on the simple substrate network, the substrate impedance Zs is represented as

$$Zs = 1/(j\omega C_{box}) + 1/(j\omega C_s + G_s) \qquad (1)$$

The substrate quality factor Qs is then calculated as

$$Qs = im(Zs)/re(Zs) \qquad (2)$$

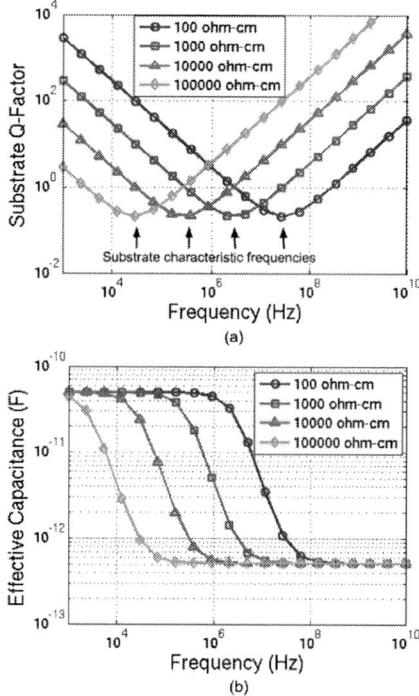

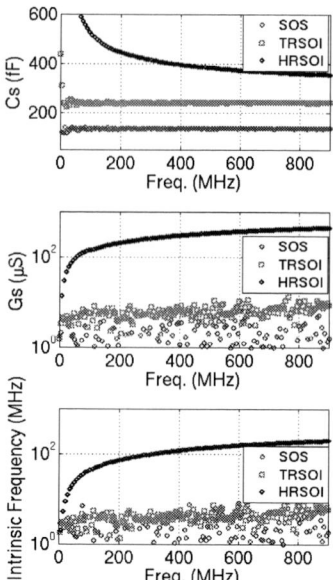

Fig. 4: Substrate capacitance, conductance and intrinsic frequencies extracted from impedance measurements of an open CPW stub.

Fig. 3: (a) Substrate quality factor versus frequency for different resistivities of handle wafer, and (b) effective substrate capacitance versus frequency for different resistivities of handle wafer.

Fig. 3 shows Qs as a function of frequency for different substrate resistivities. It is seen that the Q-curve has a minimum at what we call here the "substrate characteristic frequency", f_{sc}, where the substrate transitions from exhibiting a predominantly resistive behavior to a capacitive one. At frequencies below the characteristic frequency, the substrate capacitance is dominated by C_{box}, and is effectively higher than the high-frequency limit of C_s, as shown in Fig. 3(b).

At low frequencies, the reactance of C_s is high enough so that the substrate is effectively the series connection of C_{box} and G_s, and the series Q is given by $G_s/\omega C_{box}$. At high frequencies, the reactance of C_{box} is low enough so that the substrate is effectively the parallel connection of C_s and G_s, and the shunt Q is represented as $\omega C_s/G_s$. The substrate characteristic frequency is calculated by equating the series and shunt Q-factors of the substrate, resulting in

$$f_{sc} = f_i \sqrt{\frac{C_s}{C_{box}}} \qquad (3)$$

where

$$f_i = G_s/(2\pi C_s) \qquad (4)$$

is defined as the intrinsic frequency of the handle wafer material. By extracting the frequency response of various substrates, characteristic and intrinsic frequencies of the substrate are directly indicative of substrate resistivity and

are useful metrics for evaluating quality of SOI substrates for RF applications.

III. MEASUREMENTS AND ANALYSIS

A. Transmission Lines

To measure substrate properties, a 1-mm long CPW transmission line is fabricated on SOS, HRSOI and TRSOI substrates. The signal line is on the lowest metal level that is closest to the substrate with 25um width, and with characteristic impedances between 45Ω and 50Ω depending on the substrate. The capacitance (C_s) and conductance (G_s) of the on-wafer CPW lines are measured from 1-900 MHz in a 1-port open stub configuration using an Agilent E4991A Impedance Analyzer [Fig. 4], after short-open-load calibration to move the reference plane of measurement to the probe-tips. The intrinsic substrate frequency is calculated using (4) as shown in Fig. 4. Due to the highly insulating nature of sapphire, SOS technology has the lowest intrinsic frequency. In contrast, HRSOI has a much higher intrinsic frequency due to PSC effects at the BOX-substrate interface. The PSC effects are significantly mitigated by the introduction of the trap-rich layer in TRSOI technology resulting in a lowering of its intrinsic frequency compared to HRSOI.

Next, harmonics of the CPW lines are measured to compare linearity of the different substrates. A fundamental tone at 915 MHz is injected into one end of the transmission line with varying input power to measure the insertion loss, and the 2nd and 3rd harmonic power at the output of the transmission line. The measured results

978-1-4799-8198-4/15 $31.00 © 2015 IEEE

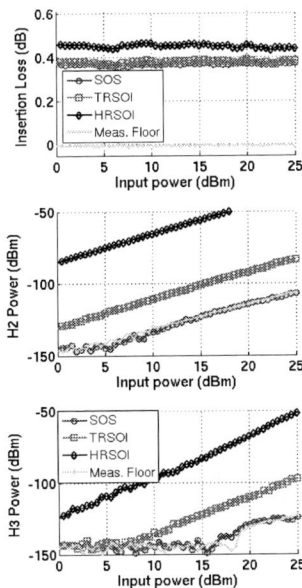

Fig. 5: Insertion loss, second harmonic power, and third harmonic power in SOI substrates measured using CPW transmission lines.

are shown in Fig. 5. It is seen that, while insertion losses of SOS and TRSOI are comparable, HRSOI losses are higher due to PSC. The metal conductivities in each process are very similar and have a negligible impact on insertion loss. In terms of linearity, SOS provides the best performance among all substrates, and is so linear that harmonic measurements are limited by the measurement instrumentation. Harmonic floor is determined by measuring a thru-line on impedance-standard substrate. TRSOI has improved linearity compared to HRSOI due to introduction of the trap-rich layer, but generates higher harmonics compared to SOS.

B. Parallel Plate Capacitors

Fig 6. (a) shows the cross-sectional view of a parallel plate capacitor built on the lowest metal layer on SOI substrate. The metal plate is wirebonded to a PCB signal line and the bottom of the SOI substrate is attached to the ground paddle of the PCB with epoxy. The ground paddle is connected to the ground plane at the bottom of the PCB through vias. Parallel plate capacitors were available only on SOS and TRSOI substrates. One-port capacitance, conductance and Q-factor measurements are taken with an Agilent E4991A impedance analyzer. The parallel plate capacitors were simulated in Ansys HFSS for the TRSOI case with the proper conductivity definitions.

Fig. 6(b) shows the measured and simulated capacitance. It is seen that TRSOI exhibits increased capacitance at lower frequencies, while SOS does not. Also, by comparing conductance measurements in Fig. 6(c) with the simulated curves in Fig. 3, SOS exhibits an effective substrate resistivity >100 kΩ-cm (measurement-limited), while TRSOI has a substantially lower resistivity

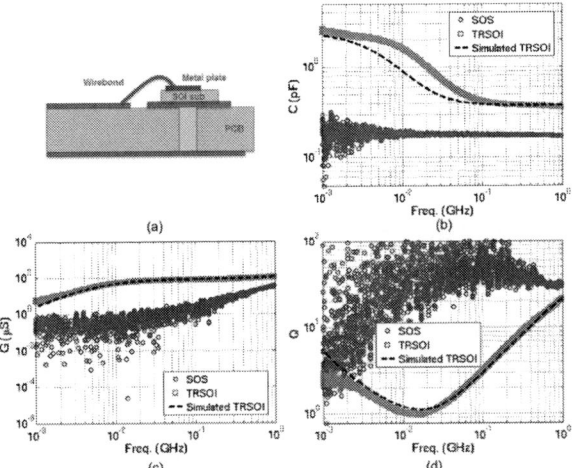

Fig. 6: (a) Cross-section of the parallel plate capacitor on SOI substrate, and measured vs. simulated (b) capacitance, (c) conductance, and (d) Q.

of a few kΩ-cm which is confirmed by full-wave EM simulations. Fig. 6(d) shows the measured and simulated quality factor versus frequency. TRSOI exhibits Q behavior that is characteristic of SOI substrates with a few kΩ-cm resistivities as seen by comparison to Fig. 3(a). The measured TRSOI substrate characteristic frequency is ~16 MHz, while that of sapphire is below the measurable frequency range. Over the measured frequency range, SOS has substantially higher Q-factor over TRSOI, which is vital to many RF applications that are sensitive to substrate losses.

IV. CONCLUSION

Comparison between various SOI technologies show that SOS is extremely linear with substrate characteristic frequency below 1 MHz while TRSOI has H2=-84 dBm, H3=-98 dBm at Pin=25 dBm, and 16 MHz substrate characteristic frequency. HRSOI provides poor RF performance due to PSC effects. For RF, SOS technology provides the best performance due to the insulating nature of sapphire, but for less stringent applications, TRSOI provides adequate performance to successfully develop RF products.

REFERENCES

[1] G. A. Garcia, R. E. Reedy and M. L. Burgener, "High quality CMOS on Thin (100nm) silicon on sapphire," *IEEE Electron Device Lett.*, vol. 9, no. 1, pp. 32-34, Jan. 1988.

[2] C. R. Neve, "Small- and large-signal characterization of trap-rich HR-Si/HR-SOI wafers for SoC applications," Ph.D dissertation, Dept. Elect. Eng., Université catholique de Louvain, Belgium, 2012.

[3] A. Botula et al., "A thin-film SOI 180nm CMOS RF switch technology," IEEE Topical Meeting on Silicon Monolithic Integrated Circuits in RF Systems, 19-21 Jan. 2009.

[4] E. H. Nicollian, and A. Goetzberger, "The Si-SiO2 interface — electrical properties as determined by the metal-insulator-silicon conductance technique," Bell System Technical Journal, vol. 46, no. 6, pp. 1055-1133, Jul. 1967.

Multitone-FM Analysis of MEMS Varactor Phase Noise Contribution in VCOs

Gerhard Kahmen[1] and Hermann Schumacher[2]

[1]Rohde & Schwarz GmbH, Muehldorfstrasse 15, 81671 Munich, Germany
[2]Institute of Electron Devices and Circuits, Ulm University, Ulm, Germany

Abstact — **Commonly, varactor contribution to VCO phase noise is assessed through Leeson's equations. Though widely applied, these equations do not show the flattening out of the SSB VCO phase noise close to the carrier. For this reason the Leeson approach only predicts the general shape of the phase noise far from the carrier, but cannot be used for precise simulations of the varactor's absolute phase noise contribution. This paper describes a multitone-FM approach which can be used to calculate the phase noise contribution of any varactor type, but is especially suitable for the assessment of close-to-carrier noise introduced by MEMS varactors exposed to thermal noise and vibrations.**

Index Terms — **Phase Noise, Close in Phase Noise, MEMS Varactor, Low Noise VCO, FM**

I. INTRODUCTION

RF VCOs are among the most important building blocks for modern communication, radar and test & measurement systems. Phase noise is one of the key characteristics of a VCO determining its applicability within an electronic system. Generally the frequency tuning of the VCO is achieved by varactor elements. Depending on the varactor technology and the circuit topology used for the VCO design this element can contribute a significant portion of the VCOs overall phase noise. Regarding linearity of the varactor CV characteristic, 1/f noise, power handling capability and tuning voltage range monolithically integrated MEMS varactors can be an attractive alternative to PN junctions or FET gate channel varactors normally used in integrated VCOs [1]. The drawback of MEMS varactors is the sensitivity to thermal noise introduced by the atmospheric molecular movement and vibrations if used in a mechanically non-stable environment. For this reason a precise calculation of the MEMS varactors' contribution to the VCOs overall phase noise, especially within a few hundred kHz from the carrier, is of large interest to enable a successful VCO Design.

In the first section of this paper the drawbacks of the phase noise calculation approach based on the extension of Leeson's equation are discussed. To overcome these drawbacks a phase noise analysis based on a Multitone-FM approach is introduced and applied to a MEMS varactor taking thermal noise and vibrations into account. Though presented for a MEMS varactor, this Multitone-FM approach can be applied to any type of Varactor.

II. CLASSICAL APPROACH OF MEMS VARACTOR PHASE NOISE CALCULATION

The common and broadly applied approach to calculate the phase noise contribution of a MEMS varactor found in the literature is based on the extension of Leeson's equation [2,3,4] using

$$\mathcal{L}(\omega_m) = \frac{K_W^2 w(\omega)^2}{2\omega_m^2} \qquad (1)$$

K_w represents the VCO tuning gain (Hz/m), $w(\omega)$ is the spectrum of the MEMS membrane displacement ($m/\sqrt{Hz}$). Taking the mechanical transfer function of the MEMS varactor into account, the power density spectrum of the membrane displacement can be expressed by

$$w_{thN}(\omega)^2 = 4k_B T b \frac{1}{k^2} \left(\frac{1}{\left[1 - \left(\frac{\omega}{\omega_0}\right)^2\right]^2 + \left(\frac{\omega}{\omega_0}\right)^2 4D^2} \right) \qquad (2)$$

The approach based on the extension of Leeson's equation given in (1) predicts the general 20dB per decade SSB phase noise roll-off, which is only valid at large offsets from the carrier. The flattening out of the phase noise close to the carrier is not predicted, the calculated phase noise power approaches infinity for $\omega_m \rightarrow 0$. For this reason the extended Leeson approach cannot be applied to accurately calculate MEMS varactor contribution to the VCO's overall phase noise.

III. THE MULTITONE FM MODULATION APPROACH

When exposed to thermal noise or vibrations the MEMS varactor membrane will be displaced leading to a frequency shift of the VCO's output signal. This effect can be described as a Multitone-FM of the VCO carrier. The atmospheric molecular movement acting on the MEMS membrane can be regarded as a white-noise-like infinite sum of individual modulation signals of equal amplitude while vibrations can be expressed by a sum of single tone modulation signals with complex phase relations. Starting from a one and two tone frequency modulation [5,6], this basic modulation theory can be extended to Multitone-FM with an infinite number of modulation signals as shown below. The multitone modulating signal can be expressed using

$$u(t) = U_0 e^{j\omega_0 t} \prod_i e^{j\beta_i \sin(\omega_i t + \varphi_i)} \qquad (3)$$

with ω_0 being the carrier frequency, ω_i the modulation signal frequency and phase φ_i. Introducing the modulation index

$$\beta_i = \frac{\Delta f}{f_i} = \frac{K_w |w(\omega_i)|}{f_i} \qquad (4)$$

and applying the first order Bessel function relation

$$e^{j\beta_n \sin(\omega_n t + \varphi_n)} = \sum_{n=-\infty}^{+\infty} J_n(\beta_n) e^{j(n\omega_n t + n\varphi_n)} \qquad (5)$$

the time domain modulation signal can be expressed as

$$u(t) = U_0 \prod_{i=1}^{k} \left[\sum_{n_i=-\infty}^{+\infty} J_{n_i}(\beta_{n_i}) \right] e^{j\omega_0 t} e^{j\left(\sum_{i=1}^{k} n_i \omega_{n_i} t + n_i \varphi_{n_i}\right)} \qquad (6)$$

A multiplication in the time domain translates into a convolution in the frequency domain giving the spectrum of the Multitone-FM signal.

$$U_{FM}(\omega) = U_0 [\delta(\omega + \omega_0) + (\omega - \omega_0)] * U_1(\omega) * \\ * U_2(\omega) \ldots * U_k(\omega) \qquad (7)$$

The spectrum of the Multitone-FM signal is the complex convolution of the spectrum of all involved single tone FM signals $U_i(\omega)$.

$$U_i(\omega) = U_i \sum_{m=-\infty}^{+\infty} J_m(\beta_m) [\delta(\omega + \omega_0 + m\omega_m e^{jm\varphi_m}) \\ + \delta(\omega - \omega_0 - m\omega_m e^{jm\varphi_m})] \qquad (8)$$

IV. MULTITONE-FM ANALYSIS OF A MEMS VARACTORS PHASE NOISE CONTRIBUTION

The resulting spectrum of the force acting on the MEMS varactor's membrane is the sum of complex forces created by thermal noise and multi tone vibration.

$$F_{tot}(\omega) = F_{Th}(\omega) + F_{Vib}(\omega) \qquad (9)$$

with the vibration forces calculated from modal mass m_{mod} and the acceleration $a(\omega)$ of the MEMS varactor.

$$F(\omega) = m_{mod} a(\omega) \qquad (10)$$

The thermal noise force is calculated using

$$F_{th}(\omega) = \sqrt{4 k_B T b} \qquad (11)$$

with the damping constant b and temperature T. Applying the MEMS varactor's complex mechanical transfer function

$$G(j\omega) = \frac{1}{k} \left(\frac{\left[1 - \left(\frac{\omega^2}{\omega_0^2} \right) \right] - j \left[\left(\frac{\omega}{\omega_0} \right) 2D \right]}{\left[1 - \left(\frac{\omega}{\omega_0} \right)^2 \right]^2 + \left[\left(\frac{\omega}{\omega_0} \right) 2D \right]^2} \right) \qquad (12)$$

results in the complex spectrum of the MEMS varactor's displacement

$$w(\omega) = G(\omega) F_{tot}(\omega) \qquad (13)$$

Inserting equation (12) and (13) into (4) yields the modulation index for the modulation signal ω_i

$$\beta_i = \frac{K_w}{\omega_i} |F_{tot}(\omega_i)| \frac{1}{k} \left(\frac{1}{\sqrt{\left[1 - \left(\frac{\omega_i}{\omega_0} \right)^2 \right]^2 + \left(\frac{\omega_i}{\omega_0} \right)^2 4D^2}} \right) \qquad (14)$$

Using equation (14) in (8) gives the spectrum of the single tone ω_i FM modulated signal. Complex convolution (7) of all single tone FM spectra (8) gives the spectrum of the MEMS varactor's phase noise contribution as a result of atmospheric thermal noise and vibrations.

V. RESULTS

Figure 1 shows a photo and the mechanical transfer function of the MEMS varactor used for the phase noise calculation.

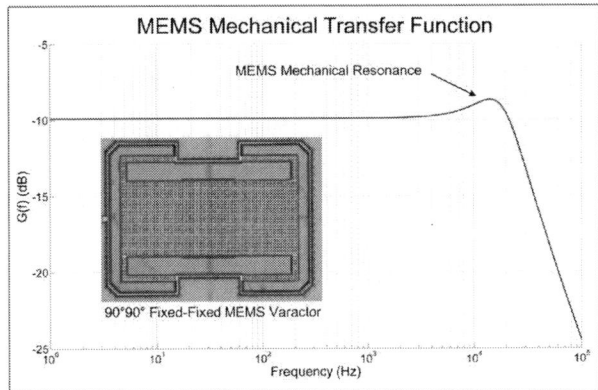

Fig. 1. Photo and simulated mechanical transfer function $G(f)$ of the Fixed-Fixed MEMS varactor used for the Multitone-FM phase noise analysis

As a first example for the Multitone-FM approach, only atmospheric thermal noise is taken into consideration for the phase noise calculation. According to equation (11) a white noise force spectrum of $1.137e\text{-}12\ N/\sqrt{Hz}$ is acting on the MEMS varactor's membrane assuming a temperature of 293°K and a damping constant of 8.0e-5 Ns/m. Applying the Multitone-FM approach for a VCO having a tuning gain of 250 MHz/µm yields the MEMS varactor's phase noise contribution shown in figure 2. This spectrum was calculated with 10 noise modulation carriers per decade in a 1 Hz resolution bandwidth. This explains the "spiky" appearance of the spectrum shown in figure 2 compared to a continuous spectrum found in measurements.

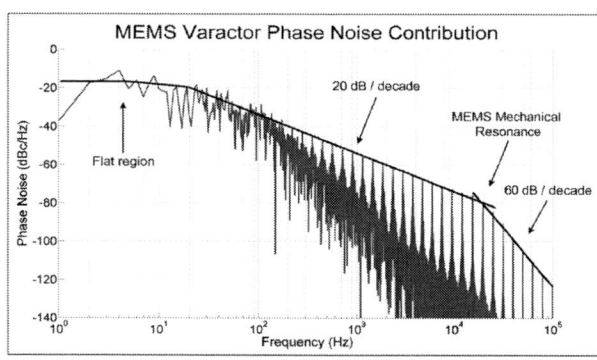

Fig. 2. MEMS varactor SSB phase noise contribution as a result of atmospheric thermal noise for a temperature of 293°K, a damping factor of 8.0E-5 Ns/m and a VCO tuning gain of 250 MHz/µm

Figure 2 shows the expected phase noise contribution of the MEMS varactor. Below the MEMS varactor's mechanical resonance frequency ω_0 a 20dB/decade roll-off and above the resonance frequency a 60dB/decade roll-off can be observed. As expected the phase noise contribution flattens out close to the carrier which can be explained by the increase of the modulation indices close to the carrier. Increasing modulation indices lead to a broadening of the single tone modulated FM spectra as a result of the Bessel function. Finally, as expected, the power of the spectrum integrated over the frequency band of interest is unity which is equal to the power of the unmodulated carrier. The unity power of the modulated carrier is a result of the Bessel functions behavior.

$$\sum_{m=-\infty}^{\infty} J_m^{\,2}(x) = 1 \qquad (15)$$

Figure 3 shows the MEMS varactor's phase noise contribution for a single 3 kHz tone vibration with an acceleration of 0.1g and a MEMS modal mass of

7.33E-10 kg additionally applied to the thermal noise above.

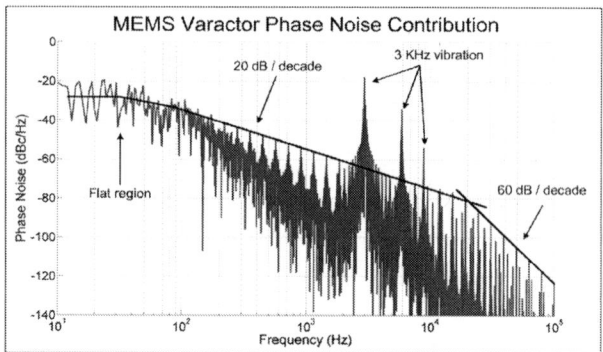

Fig. 3. MEMS varactor SSB phase noise contribution for a single 3 kHz tone vibration of 0.1g and a MEMS modal mass of 7.33E-10 kg applied additionally to the atmospheric thermal noise above

Though shown for a MEMS varactor this Multitone-FM approach is also valid for any type of electronic varactor using the noise voltage V_n instead of $w(\omega)$ in (14).

VI. CONCLUSION

The Multitone-FM approach described in this paper accurately calculates a MEMS varactor's phase noise contribution to the overall phase noise of a VCO under environmental conditions such as atmospheric thermal noise and vibrations. Contrary to Leeson's approach, this Multitone-FM analysis predicts the flattening out of the noise spectrum close to the carrier and the integrated power of the spectrum correctly. The Multitone-FM approach can be applied for any type of varactor.

REFERENCES

[1] G. Kahmen, M.Kaynak, M. Wietstruck, B. Tillack and H. Schumacher, "MEMS Varactor with High RF Power Handling Capability for Tuning of Wideband Low Noise RF VCOs", EUMC 2014, Rome, Oct. 2014
[2] Ulrich L. Rohde, Ajay K. Poddar and Georg Böck, "The Design of Modern Microwave Oscillators for Wireless Applications", John Wiley & Sons, 2005, pp. 123 - 131
[3] Gabriel M. Rebeiz, "RF MEMS: Theory, Design and Technology", John Wiley & Sons, 2003, pp. 440-443
[4] Janakiram G. Sankaranarayanan and Kartikeya Mayaram, „Noise Simulation and Modeling for MEMS Varactor Based RF VCOs", International Symposium on Circuits and Systems , pp. 2698 – 2701, New Orleans, May 2007
[5] Jens-Rainer Ohm, Hans Dieter Lüke, "Transmission of Signals", Springer, 2005, pp. 300-305
[6] Dietmar Rudolph, "FM Modulation", lecture manuscript, Technical University of Berlin

L-2L De-embedding Method with Double-T-type PAD Model for Millimeter-wave Amplifier Design

Seitaro Kawai, Korkut Kaan Tokgoz, Kenichi Okada, and Akira Matsuzawa

Dept. Physical Electronics, Tokyo Institute of Technology
2-12-1-S3-27, Ookayama, Meguro-ku, Tokyo 152-8552 Japan.
Tel: +81-3-5734-3764, Fax: +81-3-5734-3764
E-mail: kawai@ssc.pe.titech.ac.jp

Abstract—**For millimeter-wave CMOS circuit design, accurate device models are necessary. Especially an accurate de-embedding method is very important. Hence, precise de-embedding of pad parasitics is the first and valuable step to achieve accurate device models. In this work, a new pad modeling based on an L-2L de-embedding is proposed. The pad model is derived with an assumption that characteristic impedance of transmission line becomes constant at high frequency. Every device used in an amplifier is characterized with the proposed de-embedding method, and simulation and measurement results well agree with each other up to 110 GHz.**

Index Terms—**De-embedding, mm-Wave, modeling, transmission line, CMOS**

I. INTRODUCTION

Wireless communication systems in millimeter-wave (mm-wave) frequency range attracts attention from both industry and academia in order to achieve high data-rate systems. One of the best candidate mm-wave frequency range to achieve several Gbps data-rate is 60 GHz carrier, where an unlicensed 9 GHz bandwidth can be used. This 9 GHz unlicensed band enables tens of Gbps communication with a proper modulation scheme [1], [2]. For a complete TRX to be implemented with desired performances, accurate active and passive device characterization and modeling is needed, because the models provided by foundries, unfortunately, are not accurate at mm-wave frequency. Moreover, several customized devices are essential. The very first part of device characterization is the de-embedding process, which is used to remove the effects of test fixtures (probing pads' characteristics) from measured results of any kind of Test Element Group (TEG).

There are vast number of studies about de-embedding. In [3], several de-embedding methods ([4]-[6]) are discussed on the characterization and modeling of transmission lines (TLs) and their influence on amplifier response. According to [3], L-2L method is the most accurate method [6]. Still, L-2L method is not accurate enough in very high frequencies. The reason of this inaccuracy is the available less information from the measurements then required,

since in all of the mentioned methods there are two results which can be solved for two unknowns. Actually, the pad responses are not symmetrical but reciprocal. As a result, the pads should be characterized with three variables for more accurate de-embedding.

Referring to these reasons, in this paper, a new pad model is constructed with three components, and the calculation method to solve three parameters are presented with the results of L-2L de-embedding method and using the assumption of constant characteristic impedance of TLs at high frequencies.

II. CONVENTIONAL PAD MODEL

To use L-2L method, two TLs are needed, the length of one is twice the length of the other. Fig. 1 briefly illustrates the L-2L de-embedding method.

In terms of T-parameters it is expressed as follows;

$$T_{\text{meas}} = T_{\text{Lpad}} T_{\text{DUT}} T_{\text{Rpad}} \tag{1}$$

$$T_{\text{L+pad}} = T_{\text{Lpad}} T_{\text{L}} T_{\text{Rpad}} \tag{2}$$

$$T_{\text{2L+pad}} = T_{\text{Lpad}} T_{\text{2L}} T_{\text{Rpad}} \ (T_{2L} = T_L T_L) \tag{3}$$

$$T_{\text{thru}} = T_{\text{Lpad}} T_{\text{Rpad}} = T_{\text{L+pad}} T_{\text{2L+pad}}^{-1} T_{\text{L+pad}} \tag{4}$$

Here, $T_{\text{L+PAD}}$ and $T_{\text{2L+PAD}}$ are the measurement results of TLs. T_{LPAD} and T_{RPAD} are the T-parameters of left pad and right pad, respectively.

Since the measurement results of TL TEGs are symmetric and reciprocal, the thru has two known values in terms of S-, Y-, or Z-parameter. On the other hand, because pads are reciprocal passive device, they are supposed to be expressed by at least three parameters (double-T-type) pad model using three parameters shown in Fig. 2. It is impossible to determine these three parameters (Z_1, Z_2, Z_3) due to the limitations of the calculation of L-2L method, some approximation is required. Thus, only using the thru response the pads can be modeled as

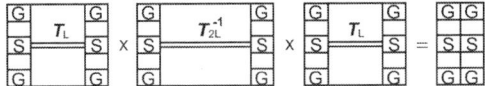

Fig. 1. Graphical illustration of L-2L de-embedding method.

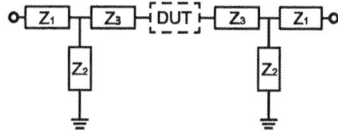

Fig. 2.　Proposed double-T-type pad model.

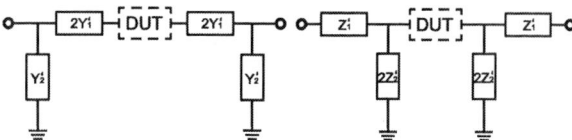

(a) π-type pad model.　　(b) T-type pad model.

Fig. 3.　Conventional pad models.

the combination of one shunt admittance and one series impedance with two different versions as π-type (Fig. 3(a)) and T-type (Fig. 3(b)) circuits. Left-pad and right-pad can be expressed by Y-parameters using π-type as follows;

$$\boldsymbol{Y}_{\text{thru\_}\pi} = \begin{bmatrix} Y_{11} & Y_{12} \\ Y_{21} & Y_{22} \end{bmatrix} = \begin{bmatrix} Y_1' + Y_2' & -Y_1' \\ -Y_1' & Y_1' + Y_2' \end{bmatrix} \quad (5)$$

$$\boldsymbol{Y}_{\text{Lpad\_}\pi} = \begin{bmatrix} Y_{11} - Y_{12} & 2Y_{12} \\ 2Y_{12} & -2Y_{12} \end{bmatrix} \boldsymbol{Y}_{\text{Rpad\_}\pi} = \begin{bmatrix} -2Y_{12} & 2Y_{12} \\ 2Y_{12} & Y_{11} - Y_{12} \end{bmatrix}$$

Similarly, T-type of thru is given in terms of Z-parameters in Eq. (6). Again in a similar way, Z-parameters of left-pad and right-pad are provided.

$$\boldsymbol{Z}_{\text{thru\_T}} = \begin{bmatrix} Z_{11} & Z_{12} \\ Z_{21} & Z_{22} \end{bmatrix} = \begin{bmatrix} Z_1' + Z_2' & Z_2' \\ Z_2' & Z_1' + Z_2' \end{bmatrix} \quad (6)$$

$$\boldsymbol{Z}_{\text{Lpad\_T}} = \begin{bmatrix} Z_{11} + Z_{12} & 2Z_{12} \\ 2Z_{12} & 2Z_{12} \end{bmatrix} \boldsymbol{Z}_{\text{Rpad\_T}} = \begin{bmatrix} 2Z_{12} & 2Z_{12} \\ 2Z_{12} & Z_{11} + Z_{12} \end{bmatrix}$$

TLs is expressed as follows in terms of F-parameter (ABCD-parameter);

$$\boldsymbol{F}_{\text{TL}} = \begin{bmatrix} \cos\gamma\ell & Z_0\sin\gamma\ell \\ \frac{1}{Z_0}\sin\gamma\ell & \cos\gamma\ell \end{bmatrix} \quad (7)$$

When double-T-type pad model is de-embedded by the π-type and T-type pad model, the de-embedded results can be expressed as following equation in terms of F-parameter.

$$\boldsymbol{F}_{\text{TL\_}\pi} = \boldsymbol{F}_{\text{Lpad\_}\pi}^{-1}\boldsymbol{F}_{\text{Lpad}}\boldsymbol{F}_{\text{TL}}\boldsymbol{F}_{\text{Rpad}}\boldsymbol{F}_{\text{Rpad\_}\pi}^{-1} \quad (8)$$

$$= \begin{bmatrix} \cos\gamma\ell & Z_0(\frac{Z_1}{Z_2}+1)^2\sin\gamma\ell \\ \frac{1}{Z_0(\frac{Z_1}{Z_2}+1)^2}\sin\gamma\ell & \cos\gamma\ell \end{bmatrix} \quad (9)$$

$$\boldsymbol{F}_{\text{TL\_T}} = \boldsymbol{F}_{\text{Lpad\_T}}^{-1}\boldsymbol{F}_{\text{Lpad}}\boldsymbol{F}_{\text{TL}}\boldsymbol{F}_{\text{Rpad}}\boldsymbol{F}_{\text{Rpad\_T}}^{-1} \quad (10)$$

$$= \begin{bmatrix} \cos\gamma\ell & \frac{Z_0}{(\frac{Z_1}{Z_2}+1)^2}\sin\gamma\ell \\ \frac{(\frac{Z_1}{Z_2}+1)^2}{Z_0}\sin\gamma\ell & \cos\gamma\ell \end{bmatrix} \quad (11)$$

Here, $\boldsymbol{F}_{\text{L,Rpad\_}\pi,\text{T}}$ are the left or right pad model of π or T-model. $\boldsymbol{F}_{\text{L,Rpad}}$ are the left and right pad model of double-T-model shown in Fig. 2.

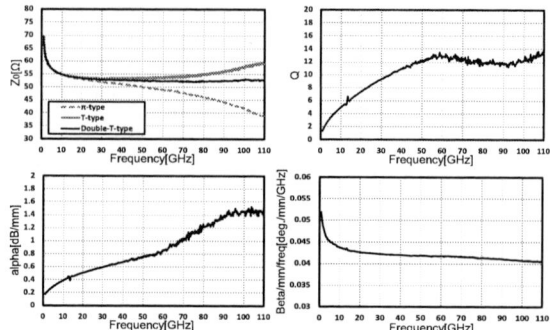

Fig. 4.　Comparison of TL characteristics.

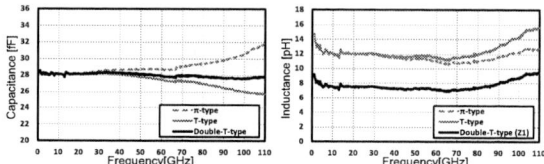

Fig. 5.　Comparison of pad models.

Thus, Q, α, and β characteristics of TLs are not different depending on the pad model but the characteristic impedance is different. When de-embedded by π-type, a $(\frac{Z_1}{Z_2}+1)^2$ fold difference occurs only in the characteristic impedance. Also a $1/(\frac{Z_1}{Z_2}+1)^2$ fold difference occurs only in the characteristic impedance by T-type pad model.

Fig. 4 compares the characteristics of the de-embedded TLs for π-type and T-type circuit of pads. It can be observed that Q, α, and β characteristics of TLs are same with each other for the two cases, but the difference between the two cases on characteristic impedance is getting larger after 20 GHz. However, it is well-known from the theory that the characteristic impedance of TL should be constant when the frequency is high, and this relation is given by the following equation;

$$Z_0 = \sqrt{\frac{R+j\omega L}{G+j\omega C}} \cong \sqrt{\frac{L}{C}}(\omega L \gg R, \omega C \gg G) \quad (12)$$

The reason of this difference can be understood by observing Fig. 5, which shows the capacitance and inductance of π- and T-type circuit of pads, accordingly. The characteristic impedance behavior and capacitance of the pad models have a direct relation. The π-model over estimates the capacitance and characteristic impedance gets smaller as the frequency increases. The counterpart of this comment can be said for T-type-model. Moreover, the capacitance of the pad resulted between top-metal and ground can be assumed constant.

III. PROPOSED PAD MODEL

As discussed in the previous section π-type or T-type pad models are expressed by two parameters (Y_1', Y_2' or Z_1', Z_2') due to limitations of the calculation of L-2L method. However, three parameters should be needed to realize the

978-1-4799-8198-4/15 $31.00 © 2015 IEEE

reciprocal passive components modeling. As mentioned, because only two parameters can be obtained from the L-2L calculation, it is necessary to put some assumption. Here, following equation is assumed:

$$Z_3 = k \times Z_1 \ (0 \le k \le 1) \tag{13}$$

From this assumption, Z_1 and Z_2 can be expressed as follow by using conventional T-type results.

$$Z_1 = Z_1' + \frac{Z_2'}{2}(k+1 - \sqrt{k^2 + 2k + 1 + 4k\frac{Z_1'}{Z_2'}}) \tag{14}$$

$$Z_2 = \frac{2k(Z_1' + Z_2')}{k - 1 + \sqrt{k^2 + 2k + 1 + 4k\frac{Z_1'}{Z_2'}}} \tag{15}$$

Then, "k" is adjusted such that the capacitor remains relatively constant up to 110 GHz. This time, k is found to be 0.4. Better to note that this value might change with different processes and different structures of pads and TLs. Fig. 5 shows the capacitance and the inductance values of the calculated proposed pad model in comparison with π-, and T-type. In the inductance of proposed pad mode, only Z_1 is included, that of Z_3 is not included. One can observe that the capacitance of the proposed pad model remains almost constant up to 110 GHz. Moreover, Fig. 4 provides Z_0, Q, α and β of characteristics of de-embedded TL by the proposed pad model. Compared to the results in Fig. 4, Q, α and β are same and the characteristic impedance become constant in the high frequency regime. One can conclude that the proposed de-embedding is more accurate than the conventional methods.

IV. EVALUATION WITH 1-STAGE AMPLIFIER RESULTS

To evaluate the accuracy of de-embedding, a 1-stage amplifier is designed, manufactured and measured. Fig. 6(a) provides the schematic of the amplifier and Fig. 6(b) shows the micro-graph of amplifier. This amplifier is constructed with TLs, bend lines, transistor (gate length of 60 nm and gate width of 2 μm by 20 fingers), MOM-capacitor (150 fF), and decoupling transmission line (Metal-Insulator-Metal Transmission Line (MIM TL)). These components are characterized after de-embedding of proposed pad model. The comparison of the measurement and modeled results of S_{21}, S_{11} and S_{22} is given in Fig. 7. S_{21} and S_{11} of the simulation results have good agreement with the measurement result. However, S_{22} of the simulation result is a bit different from the measurement result, which is caused by MIM TL modeling error. Because of the low impedance of MIM TL ($2\,\Omega \sim 3\,\Omega$), it is difficult to obtain accurate measurement results in a 50 Ω system.

V. CONCLUSION

Accuracy of conventional de-embedding methods are discussed, and it is concluded that symmetrical and reciprocal pad characteristics are not accurate enough in the

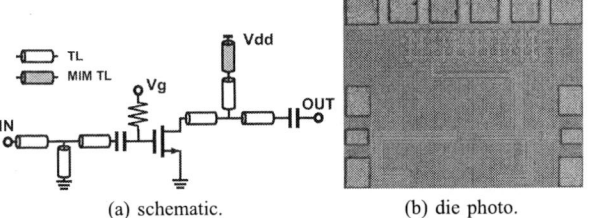

(a) schematic.   (b) die photo.

Fig. 6. 1-stage amplifier.

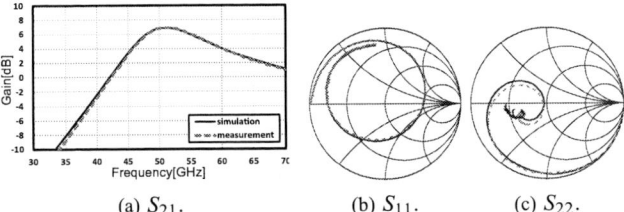

(a) S_{21}.   (b) S_{11}.   (c) S_{22}.

Fig. 7. The comparison of modeled and measured amplifier.

mm-wave range. Hence, a three element -just reciprocal-pad model is proposed along with its calculation method. The characteristic impedance of the de-embedded TLs with the proposed method remains constant in high frequencies as the theory states. Furthermore, with the new de-embedding method, several devices are characterized and its validity is evaluated with the measurement results of a 1-stage amplifier. The comparison in amplifier characteristics demonstrates accuracy of the proposed method.

ACKNOWLEDGMENT

This work is partially supported by MIC, SCOPE, MEXT, STARC, STAR and VDEC in collaboration with Cadence Design Systems, Inc., Mentor Graphics, Inc., and Agilent Technologies Japan, Ltd.

REFERENCES

[1] K. Okada, et al., "A full 4-Channel 6.3 Gb/s 60 GHz direct-conversion transceiver with low-power analog and digital baseband circuitry," in IEEE ISSCC Digest of Technical Papers, pp. 218-219, Feb. 2012.

[2] V. Vidojkovic, et al., "A low-power radio chipset in 40nm LP CMOS with beamforming for 60 GHz high-data-rate wireless communication," in IEEE ISSCC Digest of Technical Papers, pp. 236-237, Feb. 2013.

[3] R. Minami, C. Han, K. Matsushita, K. Okada, and A. Matsuzawa, "Effect of transmission line modeling using different de-embedding methods," in EuMC, pp. 381-384, Dec. 2011.

[4] M. Koolen, J. Geelen, and M. Versleijen, "An improved de-embedding technique for on-wafer high-frequency characterization," in Proc. Bipolar/BiCMOS Circuits and Tech. Meeting, Sep. 1991, pp. 188-191.

[5] H. Ito, K. Masu, "A simple through only de embedding method for on wafer s parameter measurements up to 110 GHz," in IEEE MTT-S, Jun 2008, pp. 383-386.

[6] J. Song, F. Ling, G. Flynn, W. Blood, and E. Demircan, "A de-embedding technique for interconnects," in Electrical Performance of Electronic Packaging, Oct. 2001, pp. 129-132.

[7] T. Sekiguchi, S. Amakawa, N. Ishihara, and K. Masu, "On the validity of bisection-based thru-only de-embedding," in IEEE International Conference on Microelectronic Test Structures, pp. 66-71, Mar. 2010.

Cross-Line Characterization for Capacitive Cross Coupling in Differential Millimeter-Wave CMOS Amplifiers

Korkut Kaan Tokgoz, Kimsrun Lim, Yuuki Seo, Seitarou Kawai, Kenichi Okada,
and Akira Matsuzawa

Tokyo Institute of Technology, Department of Physical Electronics, 2-12-1-S3-27
Oookayama, Meguro-ku, Tokyo, 152-8552, Japan, E-mail: korkut@ssc.pe.titech.ac.jp

Abstract— An electrically symmetric cross-line and its characterization are proposed for capacitive cross coupling in differential amplifiers. The characterization of the device is done using two structures. L-2L method is applied to achieve virtual-thru connection of Ground-Signal-Signal-Ground (GSSG) pads and fixtures used in cross-line characterization structures. Pad parasitics are modeled with T-model which provides more accurate results than Π-model. Characterization of cross-line is done using one structure and verified with the other. Comparisons show well aggrement in terms of four-port S-parameter responses up to 67 GHz.

Index Terms— Capacitive cross coupling, characterization, CMOS, cross-line, differential, mm-wave.

I. INTRODUCTION

The unlicenced 9 GHz bandwidth around 60 GHz frequency region has enabled high-data-rate wireless transceivers. Furthermore, CMOS process has several advantages compared to compound semiconductor processes [1]. Use of single-ended, or differential circuits are possible for a transceiver system. For instance, in [1] power amplifier and low-noise amplifier are designed as single-ended to decrease area, increase flexibility on layout. On the other hand, differential amplifiers are also used for RF amplifiers, local oscillator buffers, especially considering transceivers with complex modulation schemes. Moreover, fully differential transceiver architecture can also be implemented [2]. Differential amplifiers in these applications have vital roles, and neutralization techniques can be used for higher gain, low power consumption, and decreased area [1], [2]. Capacitive cross coupled amplifier is a topology for neutralization technique, for which an illustration is given in Fig. 1. Symmetry for differential amplifiers is important for decreased phase and amplitude imbalance between the two differential signals, and decreased mode conversions between common and differential modes. Although the schematic shown in Fig. 1 is symmetrical, in the layout the crossing part could be asymmetrical. Due to these reasons, a cross-line is designed for capacitive cross coupled differential amplifier to decrease phase and amplitude imbalance as in Fig. 2. In Section II, this structure is explained with the structures used for characterization of this device. Section III describes virtual-thru de-embedding method for GSSG pad

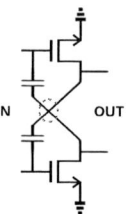

Fig. 1. An illustration for capacitive cross coupled differential amplifier (asymmetrical crossing in red dashed circle).

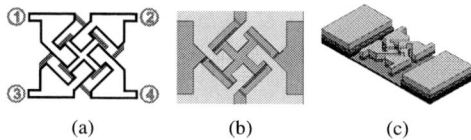

Fig. 2. Cross-line structure (a) port numbering, (b) top view (green areas are top andd gray areas are lower metal, orange areas are first two metal layers for ground), and (c) bird-eye-view.

characterization. Section IV presents experimental results.

II. CROSS-LINE AND CHARACTERIZATION STRUCTURES

Fig. 2(a) illustrates the cross-line. White areas present top metal layer and lower metal layer connected with vias. Moreover, gray areas present the lower metal layer connecting port 1 to 4 and port 2 to 3. Note that the structure is not fully symmetric considering physical layout, however, it is almost electrically symmetrical. Due to symmetry and reciprocity propoerties of the structure, the simplified S-parameters can be given as;

$$S_{\text{CCC}} = \begin{bmatrix} S_{11} & S_{12} & S_{13} & S_{14} \\ S_{12} & S_{11} & S_{14} & S_{13} \\ S_{13} & S_{14} & S_{11} & S_{12} \\ S_{14} & S_{13} & S_{12} & S_{11} \end{bmatrix} \quad (1)$$

One can observe that cross-line can be described using four S-parameters. Hence, two characterization structures are implemented. A general representation for the two characterization structures is provided in Fig. 3(a). Leftmost

978-1-4799-8198-4/15 $31.00 © 2015 IEEE

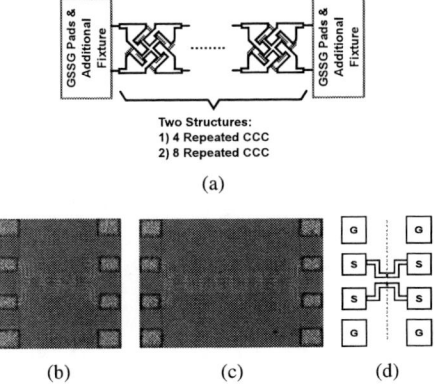

(a)

(b)　　　　　(c)　　　　　(d)

Fig. 3. Characterization structures (a) a general representation, (b) four cross-line repeated chip photo, (c) eight repeated chip photo, (d) virtual connection of fixtures.

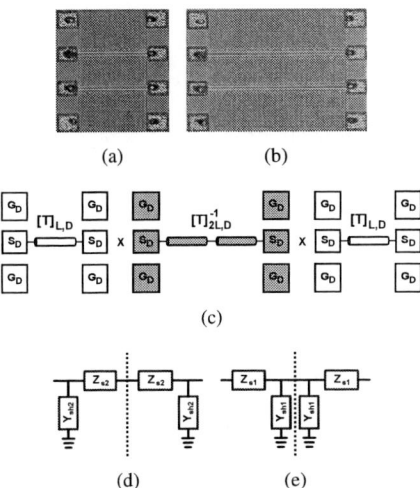

(a)　　　　　　　　(b)

(c)

(d)　　　　　(e)

Fig. 4. Virtual-thru GSSG de-embedding (a) network with 200 µm CPWs, (b) network with 400 µm CPWs, (c) virtual-thru differential mode extraction (same is done for common mode), (d) Π-model for pads, and (e) T-model for pads.

and rightmost cross-line for both structure are connected to GSSG pads with fixtures based on CPWs. In one characterization structure, four cross-line are connected to each other, and eight in the other, to decrease errors from de-embedding of pads and fixtures. The values of four and eight are selected to easily calculate only the fixture effects using virtual-Thru (L-2L) of left and right fixtures with pads are shown in Fig. 3(d). To avoid four-port T-parameter based calculations and lessen the burden, mixed-mode S-parameters are used. Mixed-mode S-parameters can be calculated in general as;

$$[S_{\mathrm{MM}}] = [M][S][M]^{-1} = \begin{bmatrix} S_{\mathrm{dd}} & S_{\mathrm{dc}} \\ S_{\mathrm{cd}} & S_{\mathrm{cc}} \end{bmatrix} \quad (2)$$

where $[S_{\mathrm{MM}}]$ is the mixed-mode S-parameter representation of a four-port network having $[S]$, S_{dd} and S_{cc} are pure differential and common modes, S_{dc} and S_{cd} are the two conversion modes. Moreover, $[M]$ is given in the following equation (I is two by two identity matrix).

$$[M] = \frac{1}{\sqrt{2}} \begin{bmatrix} I & -I \\ I & I \end{bmatrix} \quad (3)$$

Since the four-port networks used in here have symmetry and reciprocity properties, theoretically there is no conversion modes. Thus, one can divide mixed-mode of the network into two seperate two-port networks. T-parameters of four and eight repeated crossing structure for differential mode can be written as;

$$[T]_{\mathrm{4U,dd}} = [T]_{\mathrm{LP,dd}}[T]_{\mathrm{F,dd}}[T]_{\mathrm{C,dd}}^{4}[T]_{\mathrm{F,dd}}[T]_{\mathrm{RP,dd}} \quad (4)$$

$$[T]_{\mathrm{8U,dd}} = [T]_{\mathrm{LP,dd}}[T]_{\mathrm{F,dd}}[T]_{\mathrm{C,dd}}^{8}[T]_{\mathrm{F,dd}}[T]_{\mathrm{RP,dd}} \quad (5)$$

The subscripts LP, RP, F, and C are for left-, right-pad, fixture CPWs, and cross-line. By using the following equation differential response of virtual-thru connection of fixtures can be calculated as;

$$[T]_{\mathrm{PF,dd}} = [T]_{\mathrm{4U,dd}}[T]_{\mathrm{8U,dd}}^{-1}[T]_{\mathrm{4U,dd}} \quad (6)$$

$$[T]_{\mathrm{PF,dd}} = [T]_{\mathrm{LP,dd}}[T]_{\mathrm{F,dd}}[T]_{\mathrm{F,dd}}[T]_{\mathrm{RP,dd}} \quad (7)$$

where subscript PF is related with left-pad, fixtures and right-pad combinations. One can calculate the common mode of the fixtures with pads responses using the same set of equations (replace "dd" with "cc"). To calculate cross-line one needs to de-embed the pad parasitics and fixture effects. The fixture effects can be solved assuming symmetry property for fixtures after de-embedding pad responses in both pure modes.

III. VIRTUAL-THRU GSSG DE-EMBEDDING METHOD

In [3], simple thru de-embedding method is presented for four-port networks. However, this method is affected by undesired coupling between probes. For this reason, similar to [4], virtual-thru (L-2L) method for GSSG de-embedding is established. Two four-port networks with CPWs having 200 and 400 µm lengths are established. Chip photos for these two networks are provided in Fig. 4. Four-port single-ended S-parameters are converted to mixed-mode S-parameters. Again, there is no mode conversion. These four-port networks can be divided into common and differential two-port networks. Using the method shown in Fig. 4(c), one can calculate virtual connection of left and right pad in terms of differential response. Similarly common mode response can be obtained. Using virtual responses of both modes, pad parasitics for two different modes can be assumed Π- (Fig. 4(d)), or T-model (Fig. 4(e)). After measurements are done using four-port VNA, method is followed and virtual-thru response is calculated. Pad parasitics for both modes are calculated using Y-parameters for Π pad model, and Z-parameters for T-model. It can be observed from Fig. 5(a) that T-model for

978-1-4799-8198-4/15 $31.00 © 2015 IEEE　　　47

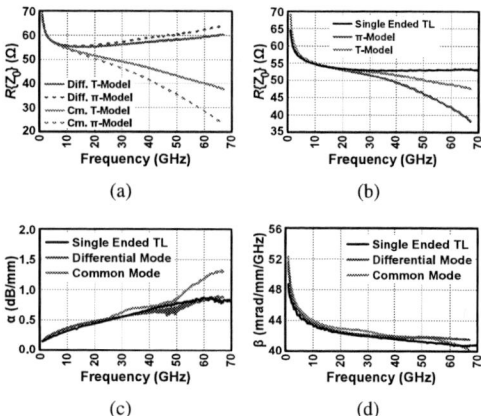

(a)　　　　　　　　(b)

(c)　　　　　　　　(d)

Fig. 5. CPW parameters for pure modes in comparison with two-port CPW characteristics (a) characteristic impedance of common (red lines) and differential (blue lines) for Π- (dashed lines), and T-model (solid lines), (b) mean values of pure modes characteristic impedance, (c) α, and (d) β.

both modes provides more accurate results than Π-model, based on the characteristic impedance response of both modes. Since there is large ground plane between CPWs, pure mode responses after de-embedding should be close to two-port CPW characteristics, this can be observed in α (Fig. 5(c)) and β (Fig. 5(d)) results after de-embedding. Both modes responses are close to each other, and also to two-port CPW characteristics obtained before this work. T-model is assumed for both modes in order to be used in de-embedding needed for cross-line characterization.

IV. EXPERIMENTAL RESULTS

From Eq. (7) pad parasitics responses can be de-embedded in terms of differential and common mode, and remaining response is two cascaded fixture response. One can solve for one fixture responses for two modes easily. To calculate cross-line results, from Eq. (4) pad parasitics and fixtures are de-embedded from left and right side. Left response is $[T]_{C,dd}^4$ for differential mode and $[T]_{C,cc}^4$ for common mode. One cross-line can be calculated easily for both modes. Four-port response of cross-line can be calculated by applying mixed-mode to single-ended S-parameter transformation. In order to verify obtained results, eight repeated structure (Fig. 3(c)) response is reconstructed in a simulation environment with calculated responses. The reconstructed model results are compared with measurements and provided in Fig. 6. Single-ended S-parameters of this four-port network has symmetry and reciprocity properties there are four different S-parameter results. It can be observed that model and measurement results well-match with each other up to 67 GHz, which validates characterized cross-line results.

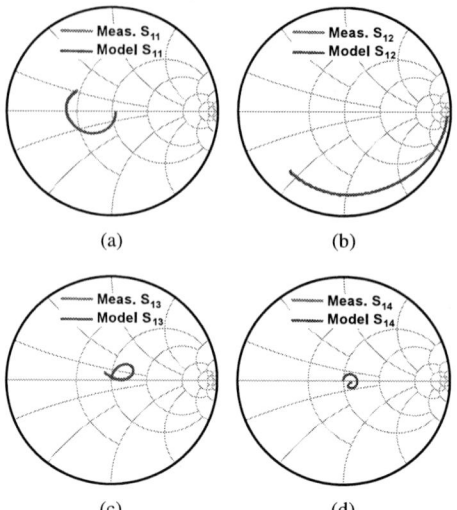

(a)　　　　　　　　(b)

(c)　　　　　　　　(d)

Fig. 6. S-parameter comparison between model (blue lines) and measurement (red lines) results of eight repeated cross-line up to 67 GHz (a) S_{11}, (a) S_{12}, (a) S_{13}, and (a) S_{14}.

V. CONCLUSION

An electrically symmetrical cross-line is introduced. Characterization of this structure is done using properties of mixed-mode S-parameters. A virtual-thru de-embedding method is used for GSSG pad parasitics calculation by applying L-2L method. This method is also applied to calculate the fixtures used in cross-line characterization structures. De-embedding of pad parasitics and fixtures are done on four repeated cross-line structure to calculate for one. Using characterized cross-line, fixtures and pad parasitics, eight repeated structure is reconstructed in a simulation environment and the results are compared with measurements. Comparisons show that there is a good match between model and measurements up to 67 GHz.

ACKNOWLEDGMENT

This work is partially supported by MIC, SCOPE, MEXT, STARC, STAR and VDEC in collaboration with Cadence Design Systems, Inc., Mentor Graphics, Inc., and Agilent Technologies Japan, Ltd.

REFERENCES

[1] K. Okada, et al., "Full Four-Channel 6.3-Gb/s 60-GHz CMOS Transceiver With Low-Power Analog and Digital Baseband Circuitry," *IEEE JSSC* , vol.48, no.1, pp.46-65, Jan. 2013.

[2] Z. Wang, et al., "A CMOS 210-GHz Fundamental Transceiver With OOK Modulation," *IEEE JSSC* , vol.49, no.3, pp.564-580, Mar. 2014.

[3] S. Amakawa, et al., "A Simple De-Embedding Method for Characterization of On-Chip Four-Port Networks," *Proc. AMC 2008*, pp.99-103, 2009.

[4] N. Li, et al., "Evaluation of a Multi-Line De-embedding Technique up to 110 GHz for Millimeter-Wave CMOS Circuit Design," *IEICE Trans. Fundam. Electron.*, Vol. E93-A, No. 2, pp.431-439, Feb. 2010.

A +18 dBm Broadband CMOS Power Amplifier RFIC with Distortion Cancellation

Ahmed M. El-Gabaly[1,2] and Carlos E. Saavedra[1]

[1]Electrical and Computer Engineering, Queen's University, Kingston, ON, Canada K7L 3N6
[2]Peraso Technologies Inc., Toronto, ON, Canada M5J 2L7

Abstract— A broadband fully-integrated class-A power amplifier (PA) is presented using derivative superposition (DS), which delivers +18 dBm of saturated output power from 1 GHz to 6 GHz. The PA exhibits a flat power gain of 13.5 ± 1.2 dB over its operating frequency range. It features a high linearity with an output 1dB compression point (OP$_{1dB}$) and third order intercept point (OIP3) of more than +16 dBm and +25 dBm respectively. The measured OIP3 remains above +25 dBm, i.e. 9 dB higher than OP$_{1dB}$, for a wide range of power levels up to 4 dB backoff from OP$_{1dB}$. The chip was fabricated using a 0.13μm CMOS process, occupying an active area of only 200μm×110μm.

Index Terms— Distortion cancellation, derivative superposition, power amplifiers, CMOS, RFIC

I. INTRODUCTION

Much of the research and development on microwave power amplifiers (PAs) is rightly focused on simultaneously maximizing their linearity and power efficiency because of the multitude of battery-operated wireless devices in our daily lives that are connected to Wi-Fi and WPAN networks. Examples of linearization techniques for amplifiers include predistortion [1]–[3], feedforward [4]–[6], derivative superposition (DS) [7]–[10], plus a few others.

Co-existing with the hand-held mobile devices there are a nontrivial number of other devices for the wireless infrastructure that draw their power from the grid. Infrastructure-related, 'non-mobile', devices include home/office routers and Wi-Fi enabled desktop computers, printers, scanners and more. As the complexity of communications standards increase in order to handle more users and data, the linearity and bandwidth specifications of RF PAs are becoming more stringent. Class A or AB PAs offer high linearity and broadband performance for grid-operated devices in comparison to switched-mode PAs (e.g. Class E or F) that are well suited for high efficiency in battery-operated mobile devices [11].

This paper describes the design and measurement of a broadband fully-integrated CMOS class-A PA with cascode transconductors that can deliver up to +18 dBm of saturated output power (P$_{SAT}$) over a wide frequency range from 1 GHz to 6 GHz. The PA exhibits a flat power gain of 13.5 ± 1.2 dB, an output 1dB compression point (OP$_{1dB}$) of +16.3 dBm, an output third-order intercept

point (OIP3) of +25.6 dBm and a power-added efficiency (PAE) of 44.3% over its operating frequency range. DS is used to reduce the distortion produced by the PA in a manner that does not require bias adjustments for different frequencies and power levels. The RF integrated circuit (RFIC) was fabricated using a 0.13-μm CMOS process and has a small active area of only 200μm×110μm.

II. CMOS PA CIRCUIT DESCRIPTION

A circuit schematic of the low-distortion CMOS PA is shown in Fig. 1 and is based on the approach used for the gallium-nitride PA described in [10]. Transistors M$_1$ and M$_2$ constitute the main signal path of the amplifier, whereas transistors M$_{1A}$ and M$_{2A}$ form the auxiliary path. M$_1$ is biased in saturation and its drain-to-source current can be written using the power series

$$i_{DS} = I_{DS} + g_{m1}v_{GS} + g_{m2}v_{GS}^2 + g_{m3}v_{GS}^3 + ... \quad (1)$$

where I_{DS} is the bias or quiescent drain-source current, v_{GS} is the gate-source voltage and

$$g_{mn} = \frac{1}{n!}\frac{\partial^n i_{DS}}{\partial v_{GS}^n}. \quad (2)$$

For strong inversion and class A operation, M$_1$ has a negative g_{m3} coefficient. M$_{1A}$ is biased near pinch-off such

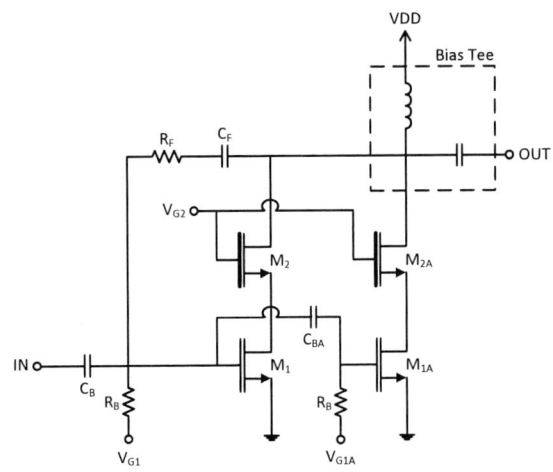

Fig. 1. Circuit schematic of the CMOS PA.

TABLE I
SUMMARY OF COMPONENT VALUES FOR THE PA

Transistor Size (μm)	$(W/L)_1$	$(W/L)_2$	$(W/L)_{1A}$	$(W/L)_{2A}$	
	200/0.12	300/0.4	75/0.12	135/0.4	
Component Values	R_B	C_B	C_{BA}	R_F	C_F
	10 kΩ	5 pF	2 pF	340 Ω	1.2 pF

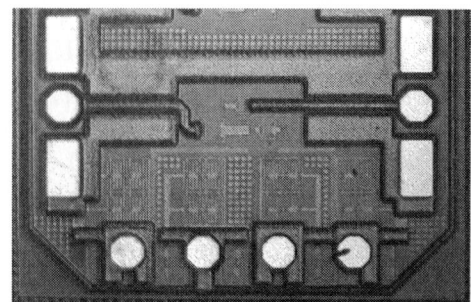

Fig. 2.   Photograph of the fabricated CMOS PA.

that its g_{m3} term is positive. In this manner, the third-order intermodulation distortion (IMD3) tones produced in the main path and the auxiliary path will have opposite phases and will cancel when they are combined at the output. The gate width and bias voltage of M_{1A} are chosen such that the IMD3 tones generated in the auxiliary path have approximately the same magnitude as those in the main path for good cancellation. Since the auxiliary path operates near pinch-off, its dc power consumption is small compared to the main one. As a result, this distortion-cancelling topology only has a minor effect on the amplifier's power efficiency.

The second-order nonlinear current of M_1, $g_{m2}v_{GS}^2$ in (1), can feed back to the gate and source of M_1 through the capacitive and inductive parasitics, generating IMD3 products which will vary with frequency. Transistor M_2 along with the gate voltage V_{G2} are designed to supply an appropriate drain voltage for M_1 such that its g_{m2} coefficient and second-order nonlinear current are close to zero. M_2 also lowers the impedance at the drain of M_1 thereby reducing the voltage swing at that node as well as the amount of (frequency-dependent) feedback to the gate of M_1 through its gate-drain parasitic capacitance C_{gd}. This improves the high frequency response and reduces the second-order harmonic signal that appears at the gate of M_1 which can mix with the fundamental tone (again through $g_{m2}v_{GS}^2$) yielding IMD3. Furthermore, the M_1-M_2 cascode allows the use of a higher supply voltage for a larger RF output swing and power level.

The amplifier uses shunt-shunt feedback through R_F and C_F for flat wideband operation and to help with impedance matching at the input and output ports. Table I summarizes the transistor gate dimensions, capacitor values and resistor values used in the design of the PA. All of these devices are integrated on-chip, using poly resistors, and metal-insulator-metal (MIM) capacitors. Transistors M_2 and M_{2A} are 3.3 V thick-oxide IO devices for reliable operation with a 2.5 V supply and large output voltage swings.

III. EXPERIMENTAL RESULTS

The proposed broadband PA was fabricated in 130 nm CMOS process and a photograph of the IC is shown in Fig 2. It occupies a total area of 0.425 mm^2 including bonding pads and decoupling capacitors, while the core circuit area is only 0.022 mm^2.

The broadband PA was measured directly on-wafer using 40GHz coplanar waveguide (CPW) probes and DC probes. An external bias tee was used at the PA output (Fig. 1) followed by an attenuator to avoid driving the spectrum analyzer and power meter at excessively high power levels.

Fig. 3 shows the measured gain, P_{SAT} and OP_{1dB} from 1 GHz to 6 GHz. The measured gain is above 14 dB up to 4 GHz, and exceeds 12 dB at 6 GHz. P_{SAT} is flat over the frequency band varying by less than 1 dB between +17 dBm and +18 dBm. The measured OP_{1dB} is higher than +15.5 dBm over the entire bandwidth, reaching a maximum value of +16.8 dBm. The mean value is +16.3 dBm and the variation is less than $\pm$0.8 dBm from 1 GHz to 6 GHz.

Two-tone measurements were also carried out for output power levels ranging from +7 dBm to +16 dBm (OP_{1dB}) over the 1 GHz to 6 GHz frequency band. Fig. 4 shows the output third-order intermodulation (IM3) distortion and OIP3 at 3 GHz, 4 GHz and 5 GHz versus the total output power (P_{OUT}). At 4 GHz, the observed IM3 is better than 40 dBc at +9 dBm and higher than 30 dBc at +13.1 dBm (i.e 3.4 dB backoff from OP_{1dB} = +16.5 dBm). The corresponding OIP3 values are higher than +26.4 dBm at +9 dBm output power and higher than +25.1 dBm at +13.1 dBm. Similar results are seen at 3 GHz and 5 GHz. Fig. 5 shows the IM3 and OIP3 at 4 dB backoff from OP_{1dB} from 1 GHz to 6 GHz. In this case, the measured IM3 and

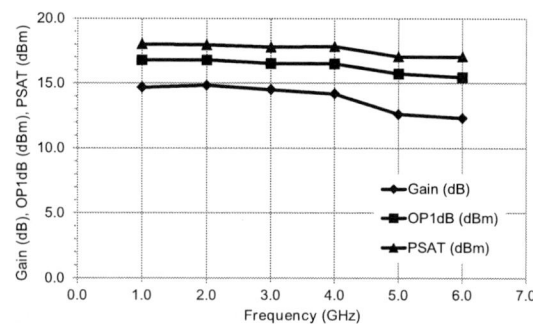

Fig. 3.   Measured gain, P_{SAT} and OP_{1dB} from 1 to 6 GHz.

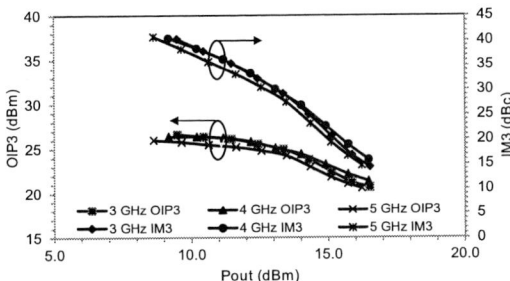

Fig. 4.   Measured IM3 and OIP3 versus P_{OUT}

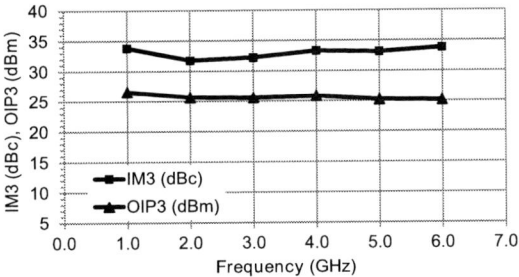

Fig. 5.   Measured IM3 and OIP3 from 1 GHz to 6 GHz.

OIP3 are better than 31.8 dBc and +25.1 dBm respectively up to 6 GHz. Overall, an IM3 better than 30 dBc and an OIP3 better than +25 dBm were achieved from 1 GHz to 6 GHz at power levels reaching +12 dBm or 4 dB backoff from OP_{1dB}.

Table II summarizes the performance of this PA with that of other broadband CMOS PAs reported in the literature [12]–[14].

IV. Conclusion

Reducing the sensitivity of the DS method to RF frequency and power level is critical for widespread industrial use in broadband applications. The DS method is advantageous for linearizing monolithic amplifiers because it requires very little additional space on-chip. In this

TABLE II
SUMMARY OF BROADBAND PA CHARACTERISTICS

Characteristic	This work	[12]	[13]	[14]
CMOS Process	**130 nm**	180 nm	90 nm	180 nm
Area (mm²)	**0.022**	0.684	0.697	0.414
Freq. (GHz)	**1–6**	1–5	5.2–13	0.1–1.2
P_{SAT} (dBm)	**18**	20–22	25.2	19.5–20.5
Gain (dB)	**13.5±1.2**	15–20	18.5	22.5±2.5
OP_{1dB} (dBm)	**16.3**	18–20	22.6	17.1–19.1
OIP3 (dBm)	**25.6**	–	see note[1]	see note[2]
PAE (%)	**44.3**	18–36	21.6	19.5–27

[1] > 25 dBc third-order harmonic distortion at OP_{1dB}.
[2] 24 dBc IM3 at an input power level of −10 dBm.

paper we demonstrated a CMOS PA that employs DS with cascode transconductors to achieve low second- and third-order distortion for high OIP3. The OIP3 performance is maintained over a broad range of frequencies and power levels without re-adjusting the bias.

ACKNOWLEDGMENT

The authors would like to acknowledge the products and services provided by CMC Microsystems, including CAD tools and chip fabrication services. Ahmed EI-Gabaly was the recipient of an NSERC Doctoral Graduate Scholarship for the period 2009 to 2011.

REFERENCES

[1] K. Onizuka, H. Ishihara, M. Hosoya, S. Saigusa, O. Watanabe, and S. Otaka, "A 1.9 GHz CMOS power amplifier with embedded linearizer to compensate AM-PM distortion," *IEEE J. Solid-State Circuits*, vol. 47, no. 8, pp. 1820–1827, Aug. 2012.

[2] J. Son, I. Kim, S. Kim, and B. Kim, "Sequential digital predistortion for two-stage envelope tracking power amplifier," *IEEE Microw. Compon. Lett.*, vol. 23, no. 11, pp. 620–622, Nov 2013.

[3] K.-Y. Kao, Y.-C. Hsu, K.-W. Chen, and K.-Y. Lin, "Phase-delay cold-FET pre-distortion linearizer for millimeter-wave CMOS power amplifiers," *IEEE Trans. Microw. Theory Techn.*, vol. 61, no. 12, pp. 4505–4519, Dec 2013.

[4] K.-J. Cho, J.-H. Kim, and S. Stapleton, "A highly efficient doherty feedforward linear power amplifier for W-CDMA base-station applications," *IEEE Trans. Microw. Theory Techn.*, vol. 53, no. 1, pp. 292–300, Jan. 2005.

[5] J. Legarda, J. Presa, E. Hernandez, H. Solar, J. Mendizabal, and J. Penaranda, "An adaptive feedforward amplifier under "maximum output" control method for UMTS downlink transmitters," *IEEE Trans. Microw. Theory Techn.*, vol. 53, no. 8, pp. 2481–2486, Aug. 2005.

[6] H. Choi, Y. Jeong, C. D. Kim, and J. Kenney, "Efficiency enhancement of feedforward amplifiers by employing a negative group-delay circuit," *IEEE Trans. Microw. Theory Techn.*, vol. 58, no. 5, pp. 1116–1125, May 2010.

[7] V. Aparin and L. Larson, "Modified derivative superposition method for linearizing FET low-noise amplifiers," *IEEE Trans. Microw. Theory Techn.*, vol. 53, no. 2, pp. 571–581, Feb. 2005.

[8] B. R. Jackson and C. E. Saavedra, "A CMOS amplifier with third-order intermodulation distortion cancellation," in *IEEE Topical Meeting Silicon Monolithic Integrated Circuits RF Systems*, Jan. 2009, pp. 1–4.

[9] J. Lee, J. Lee, B. Kim, B.-E. Kim, and C. Nguyen, "A highly linear low-noise amplifier using a wideband linearization technique with tunable multiple gated transistors," in *IEEE Radio Freq. Integr. Circuits Symp.*, June 2013, pp. 181–184.

[10] A. El-Gabaly, D. Stewart, and C. Saavedra, "2-W broadband GaN power-amplifier RFIC using the f_T doubling technique and digitally assisted distortion cancellation," *IEEE Trans. Microw. Theory Techn.*, vol. 61, no. 1, pp. 525–532, Jan 2013.

[11] Y. Song, S. Lee, E. Cho, J. Lee, and S. Nam, "A CMOS Class-E power amplifier with voltage stress relief and enhanced efficiency," *IEEE Trans. Microw. Theory Techn.*, vol. 58, no. 2, pp. 310–317, Feb 2010.

[12] P.-C. Huang, Z.-M. Tsai, K.-Y. Lin, and H. Wang, "A high-efficiency, broadband CMOS power amplifier for cognitive radio applications," *IEEE Trans. Microw. Theory Techn.*, vol. 58, no. 12, pp. 3556–3565, Dec 2010.

[13] H. Wang, C. Sideris, and A. Hajimiri, "A CMOS broadband power amplifier with a transformer-based high-order output matching network," *IEEE J. Solid-State Circuits*, vol. 45, no. 12, pp. 2709–2722, Dec 2010.

[14] H. Wu, L. Wang, P. Zhou, and J. Ma, "A 0.1–1.2 GHz CMOS ultra-broadband power amplifier," in *IEEE Int. Microw. Symp.*, June 2014, pp. 1–3.

A 1.8 to 2.4 GHz Stacked Power Amplifier Implemented in 0.25μm CMOS SOS Technology

Sultan R. Helmi, Hengying Shan, and Saeed Mohammadi

Purdue University, West Lafayette, IN, 47906, USA

Abstract -- A fully-integrated power amplifier (PA) designed with 8 stacked transistors is implemented in a 0.25μm Ultra-CMOS Silicon-on-Sapphire (SOS) technology. The stacked Cascode configuration allows the PA to deliver high gain and high output power while maintaining the PA stability. At 2.2 GHz, the PA under a bias supply of 16 V (2V per transistor) measures a saturated output power P_{SAT} of 28.5 dBm (0.7 Watt) and a linear gain of 21.7 dB with a peak power-added efficiency (*PAE*) and drain efficiency (DE) of 16% and 18.5%, respectively. *PAE* and *DE* increase to 25.5% and 29%, respectively, when the PA is biased with a 13 V power supply. In the frequency range of 1.8 to 2.4 GHz the P_{SAT} was above 27.6 dBm.

Index Terms — CMOS, SOS, Radio Frequency, stacked power amplifiers.

I. INTRODUCTION

Implementing power amplifiers (PAs) with high output power in advanced CMOS technologies is challenging due to the down scaling of the supply voltage with technology nodes, which affects the maximum saturated power of the PA. To achieve high output power, very wide multi-finger transistors in combination with differential designs or power-combining techniques have been utilized [1-3]. While wide transistors are prone to catastrophic failure due to their high un-regulated currents, power-combining techniques require large chip areas [3]. A third solution to achieve high output power is stacking of transistors. Stacking allows a high voltage supply while regulating the current and can utilize relatively narrow transistors. One of the stacking topologies is based on stacking of a common-source transistor with several common-gate transistors (multi-Cascode) to achieve high power, high gain and high output impedance [4]. The topology is limited in the number of transistors that can be stacked due to the required gate bypass capacitor values being reduced as the number of stacked transistors increases. These bypass capacitors help regulating the voltage swing across each transistor. A different stacking approach utilizes input transformers connected to each stacked cell to isolate the DC biasing of each cell from the input [5-6]. The cells can be as simple as a common source transistor or a Cascode cell or a more complicated configuration such as a multi-Cascode cell. This approach allows the stacking of a large number of transistors, limited by parasitic capacitances of

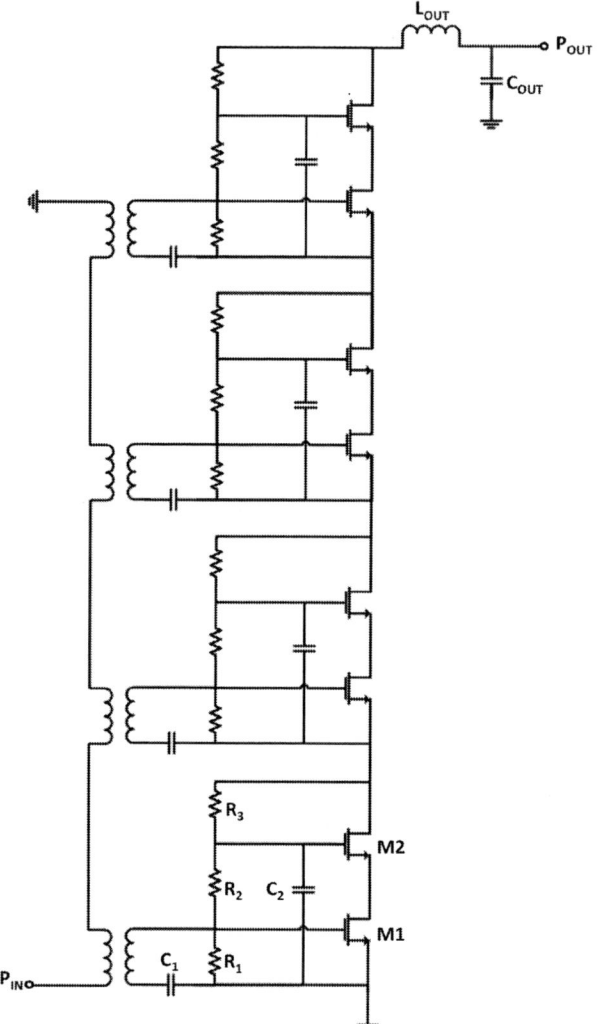

Fig. 1. The circuit schematic of the fully-integrated stacked power amplifier.

internal nodes to ground. To minimize the effect of the parasitic capacitors a substrate transfer process to a semi-insulating substrate has been proposed [5]. This approach requires an extra post-processing step. Utilizing a fully insulating substrate such as Sapphire in a CMOS Silicon on Sapphire (SOS) technology is another approach that circumvents the post-processing step [6].

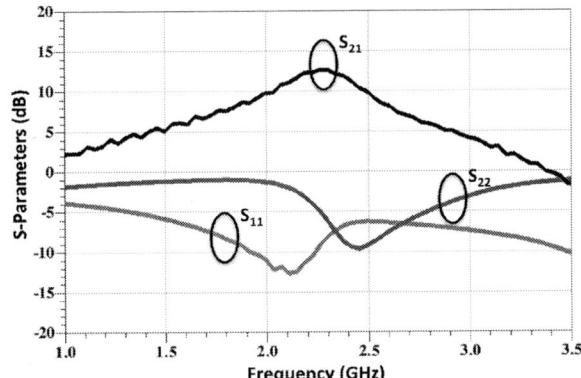

Fig. 2. Measured *S*-parameters at 12 V.

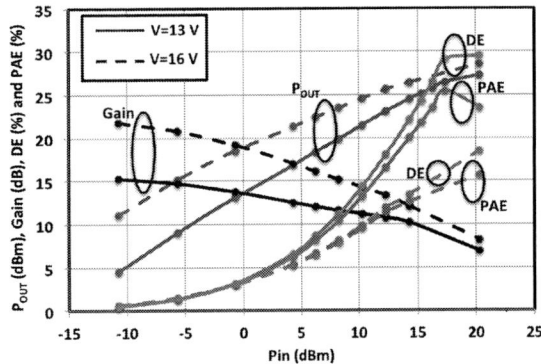

Fig. 3. Measured P_{OUT}, Gain, *PAE*, and *DE* at 2.2 GHz when biased under supply voltages of 13 V and 16 V.

In this work, a fully-integrated stacked CMOS SOS PA is designed and implemented in a 0.25μm Ultra-CMOS SOS technology. The stack configuration increases the overall voltage swing as well as the output and optimum load impedance. The PA delivers a high gain of 21.7 dB with saturated output power of 28.5 dBm at 2.2 GHz and a saturated output power above 27 dBm (0.5 Watt) across the frequency range of 1.8 to 2.4 GHz.

II. STACKED PA DESIGN

The stacked PA circuit schematic, shown in Fig. 1, is implemented in a standard 0.25μm CMOS Silicon-on-Sapphire (SOS) technology. The PA consists of 4-stacked cells with each cell consisting of a Cascode configuration. The Cascode cells provide high gain, high output impedance and better stability than their common-source transistor counterparts. Each cell is dynamically biased through a ladder of feedback resistors (R_1 to R_3), which mitigates the possibility of gate-oxide breakdown [5]. The values of capacitors C_1 and C_2 are optimized to be 8 pF each to achieve identical voltage swings across each stage leading to high output voltage swing while maintaining stable operation across the band of interest. The width of each transistor is optimized to be 7 mm with finger widths of 20 μm to achieve the required output impedance at frequencies between 1.8 to 2.4 GHz. Four input transformers are utilized to couple the input signal and provide matching to 50 Ω input impedance. To match the output a simple LC matching circuit is provided to minimize the loss of the passive components.

The design benefits from the highly insulating substrate, which reduces the internodal parasitic capacitors to ground, leading to uniform and in-phase waveforms across each cell. The stacking provides higher output impedance, which reduces the required impedance transformation ratio leading to a wide bandwidth of operation.

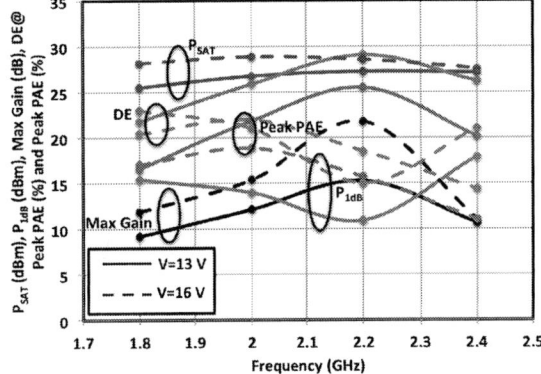

Fig. 4. Measured P_{OUT}, P_{1dB}, linear *Gain*, maximum *PAE*, and *DE* from 1.8 to 2.4 GHz when the PA is biased under supply voltages of 13 V and 16 V.

III. MEASUREMENT RESULTS

The small signal S-parameters are measured using a PNA Agilent E8361A network analyzer. The small signal S_{21} gain at the center frequency of 2.2 GHz as shown in Fig.2 is around 13 dB when the PA is biased with a 12 V power supply. Input and output reflection coefficients (S_{11} and S_{22}, respectively) dip below -10 dB with S_{21} peaking when both input and output are matched as shown in Fig.2. The large-signal performance is measured with a CW signal provided by an Agilent 83640L signal generator and an Agilent 8349A driver amplifier. The measured output power, power gain, power-added efficiency (*PAE*) and drain efficiency (*DE*) of the amplifier at 2.2 GHz under two different power supply voltages of 13 and 16 V are shown in Fig.3. Under a 13 V power supply, the PA measures a maximum saturated power P_{SAT} and a 1-dB compression point P_{1dB} of 27.2 dBm and 10.9 dBm, respectively, with a maximum linear power gain of 15.3 dB and a peak *PAE* of 25.5% with a

978-1-4799-8198-4/15 $31.00 © 2015 IEEE

TABLE I

COMPARISON OF CMOS POWER AMPLIFIERS

Reference	Technology	Frequency (GHz)	Gain (dB)	P_{SAT} (dBm)	PAE (%)	V_{DD} (V)	Area (mm²)	Topology
[1]	CMOS 0.18μm	1.95	26	26	46.4	3.4	0.832	2-stage cascode
[2]	CMOS 0.18μm	1.95	23.7	30.5	42.1	3.4	2.72	Differential
[3]	CMOS 0.18μm	2.4	22	34	34.9	3.3	3.44	4-way Power Combining
[5]	CMOS 45nm SOI	1.5-2.4 @1.8	14	30.2	23.8	15	1.2	Stacked 8-Cascodes
[6]	CMOS 0.25μm SOS	1-1.8 @1.4	10	34.4	38	16	2.16	Stacked 4-Cascodes
This work	CMOS 0.25μm SOS	1.8-2.4 @ 2.2	15.3	27.2	25.5	13	2.16	Stacked 4-Cascodes
			21.7	28.5	15.7	16		

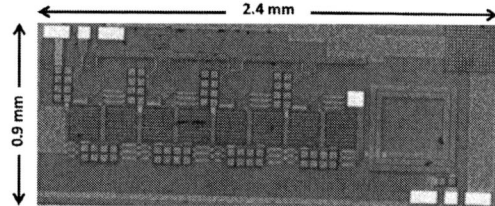

Fig. 5. Chip micrograph of the fully-integrated stacked PA.

corresponding *DE* of 29%. When biased under a higher supply voltage of 16 V at 2.2 GHz, which corresponds to 2V across each transistor (limit of safe-operating voltage), P_{SAT} and P_{1dB} increase to 28.5 dBm and 15 dBm, respectively, with a linear gain of 21.7 dB, and peak *PAE* and the corresponding *DE* of 15.7% and 18.5%. The relatively low efficiency could be related to high thermal resistance of sapphire substrate.

Fig. 4 plots the measured power performance across frequency band from 1.8 to 2.4 GHz under 13 V and 16 V supply voltages. For the supply voltage of 13 V, the power amplifier achieves P_{SAT} and maximum power gain above 25.5 dBm and 9 dB, respectively. The peak *PAE* and corresponding *DE* are above 16.5% and 21.7%. At the supply voltage of 16 V, the PA achieves P_{SAT} and maximum power gain above 27.6 dBm 11 dB, respectively. The peak *PAE* and corresponding *DE* are above 11% and 14.4%.

IV. CONCLUSION

Table I summarizes the performance of the presented PA in comparison with other reported CMOS PAs. The presented results show higher power gain compared to previously reported stacked topologies at lower frequencies while maintaining an area smaller than those based on power combining technique. The chip micrograph is shown in Fig.5, with chip dimensions of 2.4 × 0.9 mm².

ACKNOWLEDGEMENT

The authors would like to thank Peregrine Semiconductor for chip fabrication.

REFERENCES

[1] H. Jeon, O. Lee, K. H. An, Y. Yoon, H. Kim, K. W. Kobayashi, C. H. Lee, and J. S. Kenney, "A cascode feedback bias technique for linear CMOS power amplifiers in a multistage cascode topology," *IEEE Transactions on Microwave Theory and Techniques,* pp. 890-901, vol. 61, Feb. 2013.

[2] B. Koo, T. Joo, Y. Na, and S. Hong, "A fully integrated dualmode CMOS power amplifier for WCDMA applications," in IEEE Int. Solid-State Circuits Conf. Tech. Dig., Feb. 2012, pp. 82-84.

[3] J. Kim, W. Kim, H. Jeon, Y. Huang, Y. Yoon, H. Kim, C. Lee, and K.T. Kornegay, "A fully-integrated high-power linear CMOS power amplifier with a parallel-series combining transformer," IEEE J. Solid-State Circuits, vol. 47, no. 3, pp. 599-614, March 2012.

[4] Agah, J. Jayamon, P. Asbeck, J. Buckwalter, L. Larson, "A 11% PAE, 15.8-dBm two-stage 90-GHz stacked-FET power amplifier in 45-nm SOI CMOS," *Microwave Symposium Digest (IMS), 2013 IEEE MTT-S International* , vol., no., pp.1,3, 2-7 June 2013.

[5] J. Chen, S. R. Helmi, H. Pajouhi, Y. Sim, and S. Mohammadi, "A wideband RF power amplifier in 45-nmCMOS SOI technology with substrate transferred to AlN," IEEE Trans. Microw. Theory Tech., vol.60, no.12, pp.4089-4096, Dec. 2012.

[6] J. Chen, S.R. Helmi, D. Nobbe and S. Mohammadi, "A fully-integrated high power wideband power amplifier in 0.25 μm CMOS SOS technology," *Microwave Symposium Digest (IMS), 2013 IEEE MTT-S International* vol., no., pp.1,3, 2-7 June 2013.

Channelized Active Noise Elimination (CANE) With Envelope Delta Sigma Modulation

Rui Zhu, Yonghoon Song, and Yuanxun Ethan Wang

Electrical Engineering Department, UCLA, Los Angeles, California, 90095, United States

Abstract — Digital switching-mode power amplifiers (PA) based on bit-stream modulations such as Delta Sigma Modulations (DSM) have been studied as a potential solution to overcome the trade-off between the power efficiency and the signal linearity of RF PAs. FIR filtering has been introduced to suppress the quantization noise near signal band for those transmitters based on a combination of multiple power amplifiers. In this paper, the Channelized Active Noise Elimination (CANE) technique is proposed to implement the FIR filtering concept into the Envelope Delta Sigma Modulation (EDSM) scheme so that the low pass characteristics of the FIR filter is upconverted carrier to form a band pass filter. The advantage of CANE comparing to conventional FIR filtering is its small bandwidth, low sampling rate requirement and arbitrary carrier frequency and filter bandwidth selections. Simulation results are presented to demonstrate the effectiveness of the proposed approach.

Index Terms — Active noise filtering, switching-mode power amplifiers, Delta-sigma modulations, Envelope Delta-Sigma modulations

I. INTRODUCTION

Switching mode power amplifiers can achieve high power efficiency by operating in the saturation region. These PAs are well suited for the amplification of constant envelope signal but not signals with high Peak-to-Average Power Ratio (PAPR) as the linearity of modulation suffers from the saturation. Bitstream modulated RF transmitters based on Delta-Sigma modulations [1]-[6] such as Low Pass Delta-Sigma Modulation (LPDSM) or Envelope Delta-Sigma Modulation (EDSM) utilize the oversampling and noise shaping to convert non-constant envelope to constant envelope and employs an output filter to remove the out-of-band quantization noise and recycle the noise power back to DC power supply. Therefore, the high power efficiency can be obtained without corrupting the signal integrity.

However, to obtain the desired Adjacent Channel Power Ratio (ACPR), the transmitter requires either high oversampling delta-sigma modulation to shape the quantization noise further from signal band, or utilize a high quality factor output filter. Unfortunately, the former raises the complexity of the modulator design while the later usually results in bulky and lossy filter. Even though, the quantization noise near but not immediately close the signal may still locates in the filter pass band.

To solve this problem, the FIR filtering technique [8-11] is employed to suppress quantization noise in the bitstream modulated transmitter. The essential idea is to leverage on the combining effect of multiple channel power amplifiers to suppress the output noise while the high efficiency operation of power amplifier in each channel is maintained.

The FIR structure can either provide band pass filtering or low pass filtering. The bandpass filtering with FIR utilizes the higher order pass band of the filter at the signal carrier frequency to suppress the quantization noise of digitized Radio Frequency (RF) signal. LPDSM with the FIR bandpass filtering has been demonstrated by [8]. In [9] it proposes to filter the bitstreams generated by EDSM with the bandpass FIR, where the carrier frequency selection is limited by the delay time selection and broad bandwidth delay lines are required.

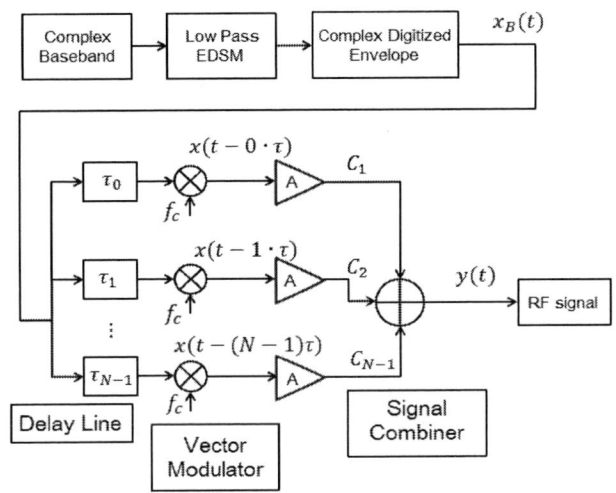

Fig. 1.   EDSM transmitter with CANE.

In [11], a polar transmitter with Delta-Sigma Modulated power supply with low pass FIR is proposed. But only the quantization noise in the envelope path is suppressed, which raises the spectrum regrowth problem in the final modulated RF signal.

To avoid the high sampling rate and the broad bandwidth delay lines, the Channelized Active Noise Elimination (CANE) technique is proposed to implement the FIR filtering concept into the Envelope Delta Sigma

Modulation (EDSM) scheme so that the low pass characteristics of the FIR filter is upconverted by carrier to form a band pass filter. Figure 1 shows an example of Envelope Delta Sigma Modulation (EDSM) with CANE.

The advantage of CANE comparing to conventional FIR filtering is its small bandwidth, low sampling rate requirement and arbitrary carrier frequency and filter bandwidth selections. For example, if a higher order pass band of the FIR filter is used for bandpass filtering, the choices of time delay unit are limited by carrier frequency to obtain a pass-band at signal band. For instance, the delay unit is selected to be $\tau = K/(2f_c)$, where K is a nonzero integer. This implies the total time delay will be much longer than the short clock period to achieve narrow band filtering, resulting in great time steps in the delay line.

In EDSM with CANE, for No.i channel, the delayed signal is

$$x_i(t) = x_B(t - i\tau)e^{j2\pi f_c t} \qquad (1)$$

Where $x_B(t)$ is the baseband complex digitized constant envelope signal.

After combining, the output signal will be

$$y(t) = Ae^{j2\pi f_c t} \sum_{i=0}^{N-1} C_i x_B(t - i\tau) \qquad (2)$$

Its Spectrum is then:

$$Y(f) = AX_B(f - f_c)H(f - f_c) \qquad (3)$$

Where $X_B(f)$ is the baseband signal spectrum and $H(f)$ is the equivalent filter frequency response, which is given by:

$$H(f) = \sum_{i=0}^{N-1} C_i e^{-j \cdot i2\pi f\tau} \qquad (4)$$

According to equation (3) and (4), the sampling rate and time delay unit are independent to carrier frequency. Thus the noise shaping performance is the only concern in selecting sampling rate. In such case, higher carrier frequency with relatively low sampling frequency is allowed, e.g., $f_c = 2$GHz while $f_s = 320$MHz. Also the time delay unit is not needed to be correlated with f_c to form a pass-band at signal carrier and it can thus be chosen arbitrarily to satisfy the filtering requirement. So the long time step required in other techniques is unnecessary in EDSM with CANE.

Note that in the EDSM transmitter structure, the phase path and digitized path are recombined by a vector modulator before being fed into switching PAs. Therefore, unlike the architecture in [11], the modulated power supplies are not needed.

II. POWER COMBINING BASED FIR FILTER

The principle of CANE relies on the power combining at the output stage of PAs. [8] uses transmission line based power combining, [9] utilizes 180 degree Hybrid blocks to combine the signals and [10][11] employ transformer based power combiner to implement the equivalent FIR filter. The turn ratio of the transformer primary and secondary coils is selected to achieve different types of filter and impedance matching.

Combining of multiple channels of signals using passive device is fundamentally lossless for the in band signal when they are essentially the same sequence even after the delay. On the other hand, the out of band noise is eliminated by either power saving mode, such as reflected back to DC supply, or power dissipating mode, such as consumed by resistive component, e.g, Wilkinson Coupler.

To demonstrate this effect, the power combining efficiency of a four channels uniform weighting 2 stages transformer based power combiner is verified in Agilent Advanced Design System (ADS), as shown in figure 2. The transformer turn ratios are selected to be 1:$\sqrt{2}$ such that all the input and output ports are matched to 50Ohm. Each channel can provide 1dBm power. The delay unit is 25ns, thus the pass-band exists periodically for every 40MHz.

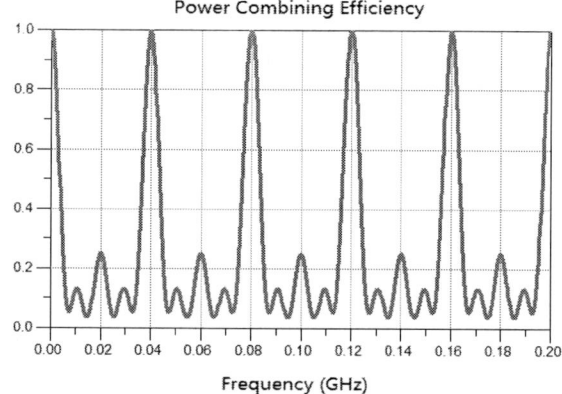

Fig. 2. 4 channels uniform weighting transformer Based Signal Combiner power combining efficiency.

In Figure 2, it shows that in band power combining efficiency is 100%. Thus, in-band signal combining of CANE is lossless. However, the output power is relatively low outside signal-band.

III. SIMULATION RESULT

The simulated spectrum of the EDSM with CANE signal is shown in figure 3 (a). It uses 4 channels uniform weighting combining, with sampling rate is 320MHz and

978-1-4799-8198-4/15 $31.00 © 2015 IEEE   56

time delay step is 8 points, corresponding to the time delay units of 25ns. The signal has a bandwidth of 3.84MHz. In the simulation of Figure 3(a), the Signal to Quantization Noise Ratio(SQNR) of EDSM CANE increases by 6dB.

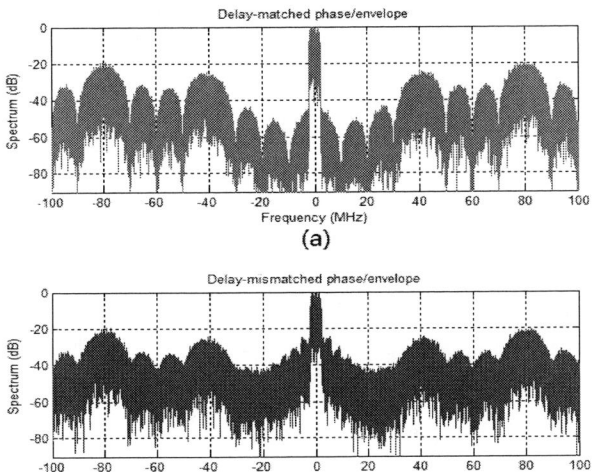

Fig. 3. (a) EDSM CANE with Delay-matched phase and envelope. (b) Noise Elimination only for Envelope without phase path delay [11].

Note that in order to indeed form the filter, one should combine the multiple channels with both phase and envelope being delayed coherently. Although only combing the digitized envelope channels [11] can eliminate the quantization noise of the envelope, the convolution of phase path and the delay-mismatched noise residue leads to spectrum regrowth, resulting in low ACPR. Figure 3 (b) shows the spectrum generated by technique in [11], with 4 channels uniform weighting combining. It is observed that the noise level near the signal band is higher in figure 3(b) than in figure 3(a).

V. CONCLUSION

From the simulation result, one can observe that the EDSM with CANE can successfully suppress the near band quantization noise of bitstream modulated transmitter. This structure decouples the carrier frequency with the sampling rate and time delay units, leading to a more flexible active filtering behavior. And the multiple channels are combined power efficiently in pass-band and the out of band noise power is eliminated.

REFERENCES

[1] A. Jayaraman, P.F. Chen, G. Hanington, L. Larson and P. Asbeck, "Linear high efficiency microwave power amplifiers using bandpass delta-sigma modulators," *IEEE Microwave and Guided Wave Letters*, vol. 8, no.3, pp.121-123, Aug. 1998..

[2] Y. Wang, "An improved Kahn transmitter architecture based on delta sigma modulation," *IEEE MTT-S Int. Microwave Symp. Dig.*, Jun. 2003, pp. 1327–1330.

[3] A. Dupuy and Y. E. Wang, "High efficiency power transmitter based on envelope delta-sigma modulation (EDSM)," *IEEE Vehicular Technology Conference*, 2004. VTC2004-Fall. 2004 IEEE 60th, vol. 3, pp. 2092-2095. IEEE, 2004.

[4] Tsai-Pi Hung; Rode, J.; Larson, L.E.; Asbeck, P.M. "Design of H-Bridge Class-D Power Amplifiers for Digital Pulse Modulation Transmitters", *IEEE Trans. on Microwave Theory and Techniques,* pp. 2845 - 2855 Vol. 55, Issue 12, Dec. 2007

[5] M. Tanio, S. Hori, M. Hayakawa, N. Tawa, K. Motoi, and K. Kunihiro, "A linear and efficient 1-bit digital transmitter with envelope delta-sigma modulation for 700MHz LTE," *IEEE MTT-S Int. Microwave Symp. Dig.*, Jun. 2014.

[6] S. Liao and Y. E. Wang, "High efficiency WCDMA power amplifier with pulsed load modulation (PLM)," Solid-State Circuits, IEEE Journal of 45.10 (2010): 2030-2037..

[7] J. Jeong and Y. E. Wang, "A polar delta-sigma modulation (PDSM) scheme for high efficiency wireless transmitters," in 2007 IEEE/MTT-S International Microwave Symposium, pp. 73-76. 2007.

[8] A. Flament; A. Frappé; A. Kaiser; B. Stefanelli; A. Cathelin and H. Ezzeddine, "A 1.2 GHz semi-digital reconfigurable FIR bandpass filter with passive power combiner," In Proc. IEEE ESSCIRC, pp. 418-421. 2008.

[9] S. Fujioka; M. Kojima; H. Izumi; Y. Umeda and O. Takyu, "Power-amplifier-inserted transversal filter for application to pulse-density-modulation switching-mode transmitters," in Communications and Information Technologies (ISCIT), 2012 International Symposium on, pp. 239-244. IEEE, 2012.

[10] R. Bhat and H. Krishnaswamy, "A watt-level 2.4 GHz RF I/Q power DAC transmitter with integrated mixed-domain FIR filtering of quantization noise in 65 nm CMOS," In Radio Frequency Integrated Circuits Symposium, 2014 IEEE, pp. 413-416. IEEE, 2014.

[11] H.Kim; T. Copani and S. Kiaei, "A 24GHz CMOS digitally modulated polar power amplifier with embedded FIR filtering," in Ph.D. Research in Microelectronics and Electronics (PRIME), 2010 Conference, July 2010, pp. 1-4.

978-1-4799-8198-4/15 $31.00 © 2015 IEEE

A 60GHz Highly Reliable Power Amplifier with 13dBm P_{sat} 15% Peak PAE in 65nm CMOS Technology

Boris Moret[1,2], Nathalie Deltimple[1], Eric Kerherve[1], Aurélien Larie[1]
Baudouin Martineau[2], Didier Belot[2]

[1]IMS Laboratory, UMR CNRS 5218, University of Bordeaux, 33405 Talence Cedex, France
[2]STMicroelectronics, Central CAD and Design Solutions (CCDS), 38920 Crolles, France

Abstract — A 60 GHz highly reliable single-ended two-stage Power Amplifier (PA) is fabricated for the Wireless Personal Area Network (WPAN) applications. The PA consists of a cascode power stage to reduce voltage drop and improving long-term reliability, and a common source driver stage. Output, inter-stage and input impedance matching networks are implemented with distributed elements (microstrip and slow-wave transmission lines). The PA achieves a saturated output power (P_{sat}) of 13dBm and a maximum Power Added Efficiency (PAE_{max}) of 15% with 13dB gain. It consumes only 84mW and occupies 0.4mm² of die area.

Index Terms — 60 GHz, millimeter-wave, power amplifier, microstrip lines, slow-wave transmission lines, cascode topology, 65nm CMOS technology.

I. INTRODUCTION

The unlicensed 7GHz band around 60GHz fulfills the multi-Gbps wireless data transfer standards for indoor WPAN applications. The main challenge for the implementation of integrated millimeter-waves (MMW) communications systems with a low cost technology is the design of efficient power amplifiers (PAs) with high output power on silicon, capable of being competitive against III/V technologies. Moreover, the low breakdown voltage of advanced CMOS technologies limits drastically the output power.

The challenge of the MMW CMOS PA design relies on the ability of delivering a maximum output power (P_{out}) with a maximum Power Added Efficiency (PAE). Unfortunately, to reach high output power, amplifiers feature high peak voltages and currents that seriously stress MOS devices. To guarantee safe device operating conditions, the supply voltage is usually less than the maximum recommended, at the price of efficiency degradation. Most of the PAs in the literature are class-A PAs in order to reach high power gain values. To reach higher gain and saturated output power, topologies that use several Power Amplifiers in parallel with Distributed Active Transformers (DAT) or power combiner can be used [1]. Hence, the latest results on MMW CMOS with SOI technology [2] or power combining structure [8] have shown 25.7 and 23.4% of PAE respectively.

In this paper, the design of a highly-reliable single-ended power amplifier is presented in order to reach a tradeoff between reliability and performance. To do so, the PA is composed of a cascode power stage, a Slow-Wave Transmission Line inter-stage and a common source driver stage. Schematic and layout are depicted in section II. Measurement results done from 57 to 66GHz are shown in the section III. It is implemented in the low cost LP (Low Power) LVT (Low Voltage Threshold) CMOS 65 nm technology from STMicroelectronics.

II. CIRCUIT TOPOLOGY

The PA structure is composed of two stages with input, inter-stage and output matching networks. Microstrip (MSTL) and Slow-Wave Transmission Lines (SWTL) are used to realize matching networks.

A. Power Stage

The Power Stage structure is shown in Fig. 1. A cascode topology is used to reduce voltage drop across the gate oxide of M1 and M2. To achieve a high current gain and a higher F_{max}, transistors' widths are determined to be 168µm. Moreover, according to [5], the optimal gate finger is 1.2µm, so 140 gate fingers are needed.

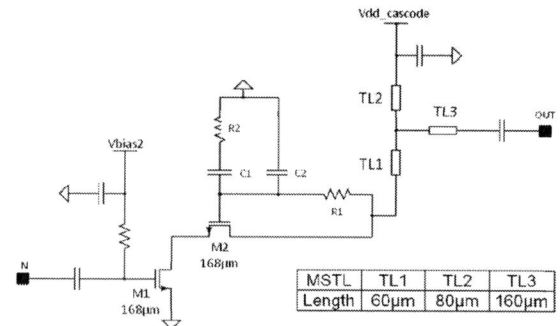

Fig. 1.   Schematic of the Power Stage.

TL1, TL2 and TL3 (Fig. 1) provide the optimal impedance at the drain of M2. The output matching network is optimized for a maximum output power and a minimum insertion loss.

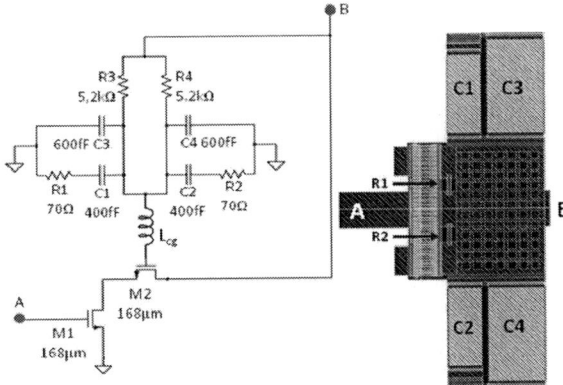

Fig. 2. Schematic with the parasitic L_{cg} and layout of the cascode.

The most straightforward way to improve efficiency is to maximize the output stage supply voltage. However device stress increases and finally a maximum *Vdd* is set to guarantee power amplifier reliability. Therefore, to insure the reliability for Time-Dependant-Dielectric-Breakdown (TDDB) and Hot-Carrier-Injection (HCI) degradation, the power stage is fed with *Vdd_cascode*=1.5V [5] instead of 1.8V usually used for cascode supply in this technology.

The biasing voltage of the common source device is chosen at *Vbias2*=0.8V then the power stage operates in class A. The biasing voltage of the common gate amplifier is chosen equal to the supply voltage to reach maximum gain with a self-biased configuration.

To mitigate the parasitic inductance L_{cg} (Fig.2) that can generate PA instability [6], interconnections are distributed under TL1 with the Metal 1-2 and the MOM capacitors (C1, C3) and (C2, C4) are divided into two equal parts with series resistors (R1, R2). Hence, the quality factor Q is reduced.

B. Inter-Stage Matching Network

An inter-stage matching network is designed for a maximum power transfer between both stages. Here instead of using a classical TL, a Slow-Wave /4 TL at 60GHz is used to reduce the physical length [3], as show in the Table I.

TABLE I
PERFORMANCES OF 65NM CMOS TRANSMISSION LINES

	MSTL	SWTL	SWTL Schematic
Z_c [•]	50	30	
[dB/mm]	1.2	2.2	
•_eff	4.2	29.4	
Q	9	13.5	
Loss_/4 [dB]	0.75	0.5	
/4 [µm]	620	230	

This line exhibits an effective permittivity of 29.4, which corresponds to a quarter-wave length of 230µm, instead of 620µm for a classical /4 MSTL in 65nm CMOS technology. The attenuation constant achieves 2.2dB/mm at 60GHz for a quality factor of 13.5. Physical length and insertion loss are drastically reduced by the use of SWTL. The /4 SWTL used has only 0.5dB of loss instead of 0.75dB for a classical /4 MSTL.

C. Driver Stage

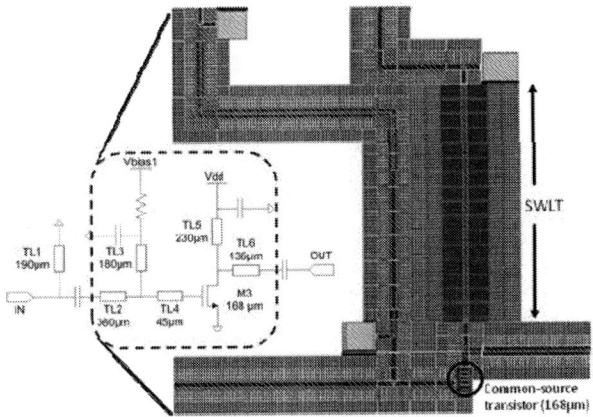

Fig. 3. Schematic and layout of the Driver Stage.

The common source driver stage operates in class AB (*Vbias1*, *Vdd*) = (0.6V, 1.2V). The transistor size is the same as the power stage ones. The input matching network is designed to maximize the small-signal gain (Fig. 3). It is composed of distributed elements. This stage is biased through the transmission lines (TL2, TL3 and TL4) providing the impedance matching.

III. EXPERIMENTAL RESULT

The proposed PA is fabricated in the 1.2V 65nm LP 1P7M CMOS process from STMicroelectronics. It enables a compact area of 1.150x0.578 mm² with DC and RF pads and 1x0.4mm² without pads. Measurements are performed on wafer at 50 input-output impedance. The PA consumes a total DC power of 84mW and 105mW for 1.5V and 1.8V cascode voltage respectively.

A. Small Signal Measurements

The S-parameters are measured thanks to an Agilent E8361A vector network analyzer and are plotted in Fig. 4. The small-signal gain is greater than 10dB over 13GHz of bandwidth (from 47GHz to 60GHz) with a maximum of 12.7dB at 55GHz. The measured S_{11} is better than -13dB at 57GHz and remains under -5dB from 42.5GHz to 61.5GHz. The measured S_{22} is less than -10dB in the whole measured bandwidth (from 40GHz to 70GHz).

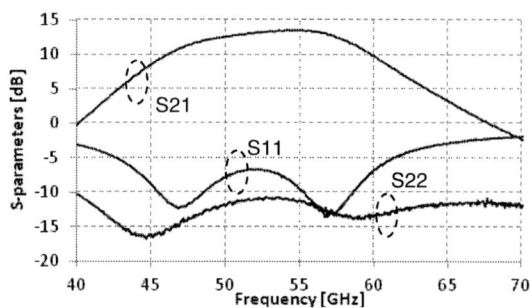

Fig. 4. Measured S-parameters

B. Large Signal Measurement

The 57GHz large-signal measurements are shown in Fig. 5. The PA achieves saturated power P_{sat} of 13dBm and 1dB-compressed P_{-1dB} output power of 9.1dBm with a peak of PAE of 15 %. When the cascode supply is moved from 1.5V to 1.8V P_{sat}, P_{-1dB} and PAE_{max} achieve 14.5dBm, 11dBm and 18% respectively.

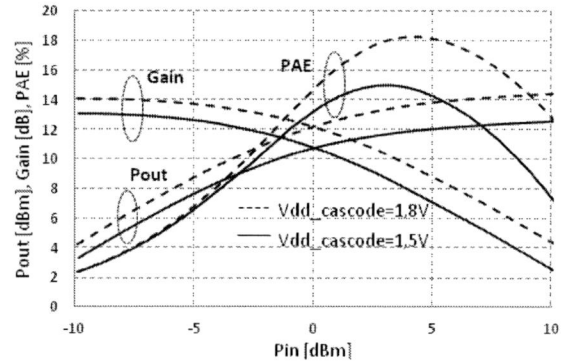

Fig. 5. 57GHz PA measured large signal P_{out}, Gain and PAE.

C. State-of-Art Comparisons

TABLE II
60GHz PAs STATE-OF-ART

Ref.	Freq GHz	Tech nm	P_{sat} dBm	P_{-1dB} dBm	PAE %	Gain dB	Pdc mW	Vdd V	Area mm²	FOM ITRS
[1]	60	40	22.6	17	7	29	2440	1.2	2.16	36425
[2]	60	65 SOI	14.5	12.7	25.7	16	77.4	1.8	0.58	1038
[3]	60	65 Bulk	13.6	9.1	9.3	14.5	185	2.4	0.16	216
[5]	60	65 Bulk	18.1	11.5	3.6	15.5	1504	1.8	0.46	302
[7]	60	90	13.8	10.3	12.6	30	178	1.8	0.33	10881
[8]	60	65 Bulk	12.9	9.3	23.4	16.7	68.4	1.2	0.11	768
This work	57	65 Bulk	13	9.1	15	13	84	1.5	0.4	194
This work	57	65 Bulk	14.5	11	18	14	105	1.8	0.4	414

FoM (ITRS) =Gain.P_{sat}.PAE.f²

Regarding to the Table I, measured power performances are comparable to the ones reported. In this work, the proposed PA achieves one of the highest reported PAE (15-18%). Highest efficiency in CMOS 65nm is obtained in [8]. Nevertheless the presented PA operates in more reliable conditions thanks to the implementation of a cascode topology than [8] that used a power combining structure with two common-source transistors.

V. CONCLUSION

A 60GHz highly-efficient two-stage PA is designed and fabricated using 65-nm Bulk CMOS technology from STMicroelectronics. Matching networks are implemented with distributed elements instead of classical lumped elements. In addition a Slow-Wave Transmission Line is used in the inter-stage matching network in order to reduce the die area and losses. A cascode power stage is implemented to propose a tradeoff between power efficiency and device reliability to improve efficiency without increasing the device stress. Finally, while operating in reliable conditions, the PA exhibits a power gain of 13dB, a saturated output power of 13dBm and a peak of PAE of 15% with a 1.5V supply voltage on the power stage. The PA demonstrates higher PAE (18%) with a 1.8V cascode voltage.

REFERENCES

[1] F. Shirinfar et al., "A fully integrated 22.6dBm mm-Wave PA in 40nm CMOS", IEEE RFIC, pp. 279-282, Jun. 2013

[2] A. Siligaris et al., "A 60 GHz Power Amplifier with 14.5dBm saturation power and 25% peak PAE in CMOS 65nm SOI", IEEE J. Solid-State Circuits, vol. 45, no. 7, pp. 1286-1294, Jul. 2010.

[3] A. Larie et al., "A compact 60GHz power amplifier using slow-wave transmission lines in 65nm CMOS," IEEE Microwave Integrated Circuits Conference (EuMIC), pp. 61-64, 2013.

[4] T. Yao et al., "Algorithmic Design of CMOS LNAs and PAs for 60-GHz Radio", IEEE J. Solid-State Circuits, vol.42, no. 5, pp. 1044-1057, May 2007.

[5] B. Martineau et al., "A 53-to-68GHz 18dBm Power Amplifier with an 8-way combiner in standard 65nm CMOS", IEEE ISSCC Dig. Tech. Papers, pp. 428-429, Feb. 2010.

[6] Y.-N. Jen et al., "Design and Analysis of a 55–71-GHz Compact and Broadband Distributed Active Transformer Power Amplifier in 90-nm CMOS Process", IEEE Transactions on Microwave Theory and Techniques, vol.57, no.7, pp.1637-1646, Jul. 2009.

[7] J.-L. Kuo et al., "A 50 to 70 GHz Power Amplifier Using 90 nm CMOS Technology", IEEE Microwave and Wireless Components Letters, vol.19, no.1, pp.45-47, Jan. 2009.

[8] W.-H. Lin et al, "A 57–66 GHz 12.9-dBm miniature power amplifier with 23.4% PAE in 65-nm CMOS", IEEE Microwave Integrated Circuits Conference (EuMIC), pp. 12-15, Oct. 2012.

RF and Microwave Technology Challenges for Internet-of-Things Applications

L. Larson

Brown University, School of Engineering, Providence, RI 02912

Abstract—An overview is provided of RF and microwave technology challenges for Internet-of-Things applications. In addition to traditional ultra-low power radio requirements, improved energy recovery and storage technologies and improved solid-state sensor technologies will be required. The marriage of RF, dc, sensors and network technologies promises a vast array of new technologies for the improvement of human health and well-being.

I. INTRODUCTION

The *Internet of Things* (*IoT*) is a catch-all phrase intended to encompass a range of emerging wireless network technologies for sensing and control on a massive and unprecedented scale. As a result, its domain overlaps many existing technologies including *Wireless Sensor Networks, RFID, Zigbee,* etc. The motivation for IoT development is the expectation that widespread, low-cost and instantaneous communication of sensor data of all types (from human health, transportation, and environmental sources), combined with "big data" analytics, will result in dramatically lower system costs and improvements in human health, safety, and overall well-being.

Widespread deployment of IoT will ultimately require improvements in RF technologies, integration of sensor technologies, improved energy storage and recovery technologies, as well as massive data mining, security, and network protocol improvements.

II. IoT SYSTEMS

In the health-care field, the IoT is generally associated with a range of technologies that unobtrusively and continuously monitor human health metrics (blood pressure, heart rate, breathing rate, blood O_2, blood CO_2, glucose levels, neural signals, troponin, etc.) and wirelessly transmit the data for clinical applications. The data can be used immediately for treatment on an individual basis, but it can also be collected from an entire patient population and used to generate radically new insights into disease prevention and cure.

This is one of the most demanding of IoT applications, because of the advanced sensor technology required, limited available power sources and the necessity to pass the stringent FDA approval process. The result has been extended development times for even some of the most promising applications.

One example of the potential of this technology is the CardioMEMS™ pulmonary blood pressure sensor, which recently gained FDA approval [1]. The CardioMEMS™ HF System provides ambulatory pulmonary artery (PA) pressure monitoring using a small pressure sensor, permanently implanted in the pulmonary artery via a catheterization procedure, often done on an outpatient basis. The sensor itself is a MEMS capacitive pressure transducer, which is resonated with an inductor; changes in the resonant frequency are a measure of the ambient pulmonary pressure, and are measured through near-field EM coupling. Patient-initiated sensor readings are wirelessly transmitted to an external electronics unit and stored in a secure website for clinicians to access and review.

Direct monitoring of pulmonary artery pressure enables early detection of worsening heart failure before noticeable symptoms appear and allows proactive management of patient care prior to a serious cardiac event [2].

In the transportation field, IoT technology is often associated with Vehicle-toVehicle (V2V) and Vehicle to Infrastructure (V2I) applications, as well as the variety of widely reported "self-driving" car technologies currently under development.

The National Highway Transportation Safety Administration (NHTSA) announced a notice of proposed rule-making in August 2014, proposing a variety of V2V technologies, all operating in the microwave band [3]. For example, widespread deployment of V2V Left Turn Assist (LTA) and Intersection Movement Assist (IMA) V2V technologies could prevent up to 600,000 crashes and save 1,000 lives per year [4]. LTA warns drivers not to turn left in front of another vehicle traveling in the opposite direction and IMA warns them if it is not safe to enter an intersection due to a high probability of colliding with one or more vehicles [4].

Widespread deployment of robust automotive sensor technologies (radar, lidar, video, etc) along with wireless communications of the resulting data is expected to reduce the fatality rate (currently still around 30,000 deaths per year in the United States). The hope here is that continuous collection of sensor data from an entire population of active cars, will allow for dramatic safety progress, as improved algorithms developed from this data are made available to all cars on a rapid basis. Studies from Google suggests that the aggregated sensor data rate from a self-driving car is on the order of 1Gb/sec – within the range of expected 5G long term wireless network improvements [5].

IoT environmental sensors could eventually be widely deployed to detect pathogens, heavy metal contamination, and explosives.

III. ENERGY RECOVERY AND STORAGE FOR IoT APPLICATIONS

Providing suitable power sources for implanted health care IoT devices is an ongoing challenge. Inductive near-field coupling is often employed, though the size of the resulting implanted transponder is often uncomfortably large.

Recently optimized mid-field coupling has been exploited to dramatically improve the efficiency of the power transfer at small implanted device scales (just a few mm on a side) [6]. In this case, a frequency of 1.9 GHz was found to provide optimum power transfer at safe exposure levels to ultra-miniature devices implanted in the heart.

The power levels of received RF energy within the body are necessarily low for health and safety reasons, and so the RF-dc conversion process is inefficient due to limits in device turn-on voltage and circuit architectures. Improvements in this conversion efficiency could result in substantial improvements in the performance of implanted devices, and this is a topic of active research in recent years [7].

The challenges associated with EM powering of implanted devices have led to an exploration of the use of the body's own power source – glucose. In this case, the oxidation of glucose can lead to power densities on the order of 180uW/cm^2, which is especially attractive for neural implant applications, since cerebral spinal fluid is largely glucose-based [8].

Another proposed power source for implanted electronic devices is ultrasonic energy, which has low losses when passed through tissue, though transduction efficiencies (sound-to-dc) can be quite poor. Again, improvements in these efficiencies could lead to significant improvement in the performance and usability of implanted devices for medical applications [9].

VII. CONCLUSION

The potential impact of widespread deployment of IoT technology over the next forty years is comparable to the impact wireless cellular technology has had over the last forty years. Improvements in RFIC technology, micro-power sources, and integrated sensors will be required in order to fully exploit the possibilities.

REFERENCES

[1] CHAMPION: Trial Rationale and Design: The long term safety and clinical efficiency of a wireless pulmonary artery pressure measuring system," P.B. Adamson, et. al, Journal of Cardiac Failure ,vol. 17, no. 1, 2011.

[2] Abraham WT, Adamson PB, Bourge RC, et al. "Wireless pulmonary artery haemodynamic monitoring in chronic heart failure: A randomized controlled trial." Lancet. 2011;377(9766);658-66.

[3] U.S. Department of Transportation Advance Notice of Proposed Rulemaking to Begin Implementation of Vehicle-to-Vehicle Communications Technology , NHTSA 34-14, August 18, 2014

[4] Vehicle-to-Vehicle Communications: Readiness of V2V Technology for Application DOT HS 812 014,

[5] http://mashable.com/2013/05/03/google-self-driving-car-sees/

[6] S. Kim, J. S. Ho, L. Y. Chen and A. S. Y. Poon, "Wireless power transfer to a cardiac implant" Appl. Phys. Lett. 101, 073701 (2012).

[7] C. Valenta, and G. Durgin, "Harvesting Wireless Power: Survey of Energy-Harvester Conversion Efficiency in Far-Field, Wireless Power Transfer Systems," Microwave Magazine, IEEE , vol.15, no.4, pp.108,120, June 2014.

[8] B. Rappport, J. Kedzierski, R. Sarpeskhar, "A Glucose Fuel Cell for Implantable Brain–Machine Interfaces," PLOS One, vol. 7, vol. 6, 2012.

[9] Seo, Dongjin, et al. "Neural dust: an ultrasonic, low power solution for chronic brain-machine interfaces." arXiv preprint arXiv:1307.2196 (2013).

978-1-4799-8198-4/15 $31.00 © 2015 IEEE

Reconfigurable Solutions for Mobile Device RF Front-ends

Arthur S. Morris III

wiSpry Inc., 20 Fairbanks, Ste. 198, Irvine, CA 92618; e-mail: art.morris@wispry.com

Abstract — Tunable RF elements are now in widespread use within mobile handsets to improve performance, compensate for environmental influences, shrink antenna volumes and shorten design cycles. The most critical challenge facing the RF front end arises from the rapidly multiplying frequency bands of operation for 4G systems around the world. In many cases, there is a multiplicity of bands even within individual countries and multiple bands/channels may be used simultaneously for carrier aggregation. Each frequency band and mode typically requires customized filtering to handle unique interference challenges. The complexity following the present design approach scales super-linearly due to the interactions between the multiple hardware chains, particularly with regards to matching and isolation. The increase in hardware also applies strong upward pressure on system area and cost while negatively impacting performance, particularly overall efficiency. We present details on development toward a fully tunable system targeted to replace the large quantity of fixed elements currently required with a scalable and compact global solution and an update on the supporting core technology.

Index Terms — tunable circuits, tunabledevices, cellular phones, radio front ends, RLC circuits, radio-frequency micro-electro-mechanical systems.

I. INTRODUCTION

Tunable RF elements are now in widespread use within mobile handsets to improve performance, compensate for environmental influences, shrink antenna volumes and shorten design cycles. Several companies have brought tuner products to the market [1-5]. As a new component in the RF designer's toolkit, the impact of tunable components on the design and implementation of modern radios is still in its infancy.

The most critical challenge facing the RF front end of modern radios arises from the rapidly multiplying frequency bands of operation for 4G systems around the world. In many cases, there is a multiplicity of bands even on single service providers within individual countries. Each frequency band and mode typically requires customized filtering to handle unique interference challenges. The complexity following the present design approach scales super-linearly with the number of bands and modes due to the interactions between the multiple hardware chains, particularly with regards to matching and isolation. The increase in hardware also applies strong upward pressure on system area and cost while negatively impacting performance, particularly overall efficiency. Additionally many of the new standards require higher levels of performance, particularly with regards to linearity. We present details on a development toward

a fully tunable system to replace the large quantity of fixed elements currently required that provides higher performance in a more compact form factor

II. HANDSET RF FRONT ENDS

Radios in modern handsets utilize an impressive array of technologies to provide outstanding overall performance at very low cost and size. Figure 1 shows a typical approach for implementing a multi-band RF front end. Acoustic filters provide sharp roll-off and low loss and can provide well over 50 dB of duplex isolation between transmit and receive frequencies. Multi-throw switches select between sets of filters and power amplifiers and provide insertion losses below 0.5 dB and isolation over 20 dB. Depending on the vendor and architecture chosen there may be individual power amplifiers per RF chain or broadband amplifiers may be utilized to cover multiple bands. Note that typically an additional chain must be added to the system for each frequency band and mode to be covered.

Not many years ago, a standard 3G handset addressed only 5 bands over 2 modes. Currently, the baseline is 20 bands over 3 modes. This is expected to increase to over 30 bands and 4 modes by 2017 with future service contemplated at 5 GHz and beyond. Further hardware proliferation arises from the separate multiband RF chains required to support MIMO and/or diversity capabilities. As implied in figure 1, the associated complexity of the RF front-end greatly increases to provide the necessary signal conditioning across all of these use cases.

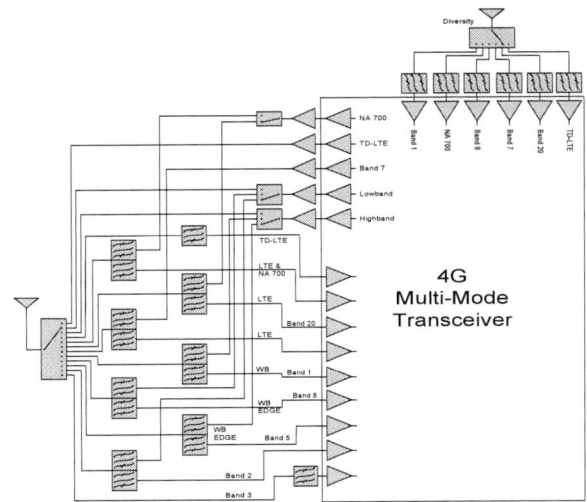

Figure 1. Conventional handset radio block diagram

Disparate technologies are used to implement conventional front-ends in handsets. The amplification is typically GaAs HBT or CMOS, the filters are acoustic utilizing BAW or SAW technologies while the multi-throw switches are based on SOI/SOS or HEMT. Hybrid integration has been employed to combine these disparate technologies, shrink the footprint of these complex systems and ease their adoption. Substantial reduction in the area per chain has also been achieved through advanced packaging techniques for the filters themselves [6-8]. However, the core of the filtering elements is difficult to further downsize due to the order of the filters needed to meet the difficult fixed selectivity requirements along with the applications' levels of power handling and linearity. Likewise, additional RF paths require more switch throws, which adds insertion loss and/or degrades isolation. We have reached a crossroads where additional capability is engendering costly area and performance penalties.

An alternative approach that has been studied for many years would be to enable a single RF chain to operate over multiple bands and modes [9-12] through reconfiguration and tuning. Past attempts at this lofty goal have not provided acceptable performance, particularly within form factors useful for handsets. However, the recent advent of higher Q and higher linearity tunable components holds the promise of enabling such flexible and scalable systems.

III. TUNING DEVICES

Successful implementation of such a system drives key component specifications including ratio, effective Q, capacitance density, voltage handling and linearity. We use RF-MEMS as our tuning devices for their outstanding performance and their ability to be monolithically integrated with CMOS for low cost and effective system partitioning. Table 1 provides specifications for one library element, a shunt cell with 1pF of capacitance shift. These and other MEMS elements are arrayed into a chip floorplan for a given application.

The implementation is based on a 0.18 μm RF-CMOS process with high voltage capability. The RF-MEMS are integrated into the back end interconnect process yielding high volume production at low cost. The CMOS design includes serial interface, power supply, and the high-voltage switches that drive the MEMS actuators. The CMOS back end is comprised of 5 metal layers including a thick metal interconnect. The MEMS are built utilizing CMOS interconnect unit processes and are sealed at the wafer level to enable standard IC packaging techniques including thinning, bumping, dicing, assembly, molding and test at standard high volume suppliers. Figure 2 shows a Wyko optical profilometer image of a fabricated 2-bit MEMS cell.

TABLE 1. Parameters of a 1pF MEMS Shunt Capacitor Cell

Parameter	Value
C_{min} (Off State)	0.13 pF
C_{max} (On State)	1.27 pF
Series Resistance (R_{series})	< 0.7 Ω
Series Inductance (L_{Series})	< 0.3 nH
Tuning Ratio	10:1
Q at C_{max} for 1 and 2 GHz	160 and 87
Pull-in Voltage	27 V
Self actuation voltage	50 Vrms
Hot switching voltage	12 Vrms
IP2	> 160 dBm
IP3	>80 dBm
2nd harmonic (Pin = +24dBm)	-110 dBm
3rd harmonic (Pin = +24dBm)	-135 dBm

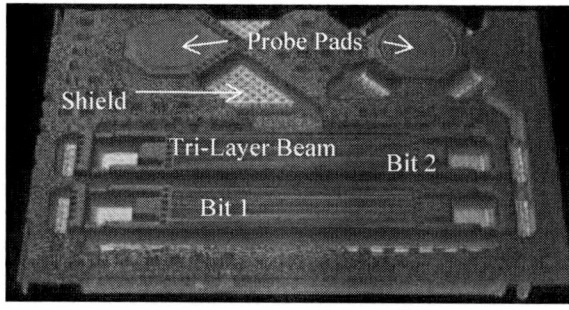

Figure 2. Profilometer image of a fabricated MEMS capacitor

Carrier aggregation is driving stringent linearity requirements in next generation systems. For example, third harmonics generated by a fundamental in Band XII will fall in the transmit band of Band IV. These third harmonics will interfere with the desired signals, and must be minimized. Other frequency combinations are equally critical, so the linearity of the tuning devices will determine the integrity of these systems. This distinction will become more visible as tunability expands in the front-end beyond antenna tuners. MEMS provides the highest linearity of any tuning technology and may be the only technology capable of supporting tuning for future systems without engendering unacceptable interference.[13]

IV. TUNABLE FRONT END CONCEPT

The key to a tunable front end is achieving the required filtering to protect the receiver from the transmitter and other interferers, while suppressing transmitter harmonic and noise emissions. The solution should also minimize insertion losses for the transmit and receive signals. Achieving both high selectivity and high efficiency requires high Q building blocks. The highest practical Q's of lumped inductive elements in the cellular bands is between 100 and 200. When used with capacitors having significantly greater Q's, resonator Q's of 100-200 are achievable. These Q's limit the filter roll-off achievable between a typical receive pass band and a transmit reject band, yielding 25-35 dB for low insertion loss designs. The additional selectivity needed in the front end can be obtained by utilizing the antennas as filters. Since the antenna occupies significant volume, the effective filter Q of the antenna can be quite high. The instantaneous bandwidth and minimum radiation efficiency of a well-designed antenna set the minimum volume that can be allocated to the antenna structure including the keep-out area surrounding the antenna structure within the handset. With careful design, the roll off in the antenna response across a typical duplex frequency spacing can easily exceed 30 dB. Thus, utilizing the combination of tunable lumped filters and tunable narrowband antennas, more than 55 dB of total duplex isolation can be achieved. The system is illustrated in figure 3 where it is assumed that the full worldwide frequency range will be addressed in 2 band ranges although 3 frequency ranges may be required for some implementations.

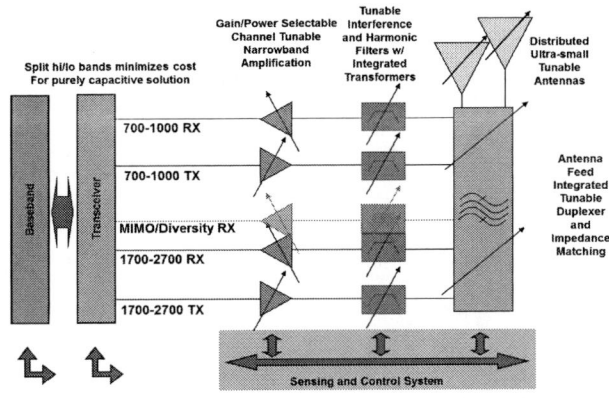

Figure 3. Block diagram of tunable radio implementation providing worldwide 4G support.

V. IMPLEMENTATION

Individual elements in the proposed system have been designed and prototyped. Transmit and receive filters have been developed for commercial mobile bands with one TX/RX filter pair covering 700 to 1000 MHz and another covering 1700 to 2700 MHz, including reverse duplex bands. These tunable receive and transmit filters are implemented using complementary asymmetric notch filters [14]. Filter measurements shown in Figure 4 demonstrate the performance required for implementation of the overall system.

The prototype tunable narrowband antennas are small monopoles with end load tuning provided by our capacitor arrays. This approach provides multiple benefits, including ultra-

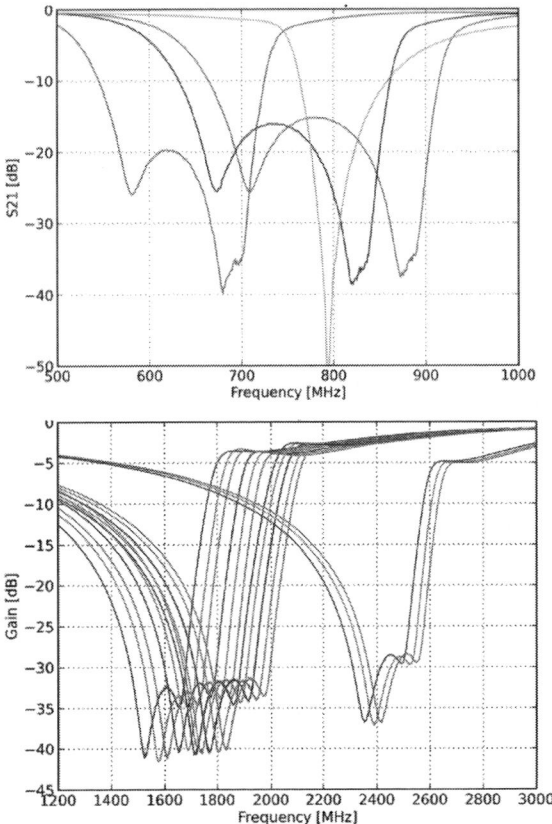

Figure 4. Tunable Asymmetrical Notch Filter Measured Responses

small physical size, wide tuning range and high selectivity for a given bandwidth and efficiency. These antennas were implemented on a smartphone-sized platform. Radiator lengths of 10 mm for high band and 25mm for low bands were used with total antenna volume of less than 1.8 cc. Operating efficiencies of 30-60% were measured. The antenna response is tunable for both duplex spacing and center frequency over the intended bands. Worst case TX-RX isolation of over 30dB has been demonstrated across the full frequency range.

The antennas and filters have been combined into an overall tunable front-end on a demonstration board having a handset form-factor. Measured results from this demonstrator will be presented. The next step will be to combine the front end with a commercial cellular transceiver to measure overall system performance for data and voice traffic including all blocking tests.

VII. CONCLUSION

The need for a fully tunable RF front end solution is rapidly growing driven by LTE deployments around the world. Technologies capable of providing the necessary efficiency, selectivity and linearity are just now coming to market. An agile RF front end architecture for handsets is proposed. Related components have been designed and prototyped and have yielded targeted performance. Initial measurements of the overall front end demonstrate the feasibility of the approach.

REFERENCES

[1] Boyle, K. et al..; "A self-contained adaptive antenna tuner for mobile phones", *6th European Conference on Antennas and Propagation (EUCAP)*, 2012, Page(s): 1804 - 1808

[2] P. McIntosh, "Harnessing the Benefits of Adaptive RF Tuning", *Wireless Design and Development*, Aug. 2008.

[3] Whatley, R.; Ranta, T.; Kelly, D.; "RF Front-End Tunability for LTE Handset Applications," *Compound Semiconductor Integrated Circuit Symposium (CSICS)*, 2010, pp.1-4, Oct. 2010

[4] S. Natarajan, et al., "CMOS Integrated Digital RF MEMS Capacitors", *IEEE 11th Topical Meeting on Silicon Monolithic Integrated Circuits in RF Systems (SiRF)*, pp 173, Jan. 2011

[5] Costa, Julio C. et al.; "SOI technology with above-IC MEMS integration for front end wireless applications", *2008 IEEE Bipolar/BiCMOS Circuits and Technology Meeting*, pp. 204 - 207

[6] R. Ruby et al., "High-Q FBAR Filter in a Wafer-Level, ChipScale Package", *ISSCC* Feb. 2002, #11-3

[7] M. Franosch et al., "A Wafer-Level-Process using Photo-Epoxy to create Air-Cavities for Bulk-Acoustic-Wave RF-Filters", *IMAPS 2004* , Long Beach California, Nov. 2004

[8] P. Carson and S. Brown, "Less is More: The New Mobile RF Front-End", Microwave Journal, June 2013, pp. 24-34.

[9] Mitola, Joseph, "Technical challenges in the globalization of software radio", *IEEE Communications Magazine*, Vol.37, No.2, Page(s): 84 - 89

[10] MacLeod, J. et al., "Enabling technologies for software defined radio transceivers" *MILCOM 2002. Proceedings*, Oct. 2002

[11] Rebeiz, Gabriel, "RF MEMS: Theory, Design, and Technology", Wiley-Interscience 2002, p 384.

[12] Okazaki, Hiroshi et al., "Reconfigurable RF Circuits for Future Band-Free Mobile Terminals", *Signals, Systems and Electronics, 2007. ISSSE '07, pp.* 99 - 102

[13] L. Dussopt and G.M. Rebeiz, "Intermodulation distortion and power handling in RF MEMS switches, varactors and tunable filters", *IEEE Trans MTT 51 (2003)*, pp 1247–1250

[14] De Luis, J. R. et al., "A tunable asymmetric notch filter using RF-MEMS", *2010 Microwave Symposium Digest*, pp. 1146 – 1149

An Integrated Reconfigurable Tuner in 45nm CMOS SOI Technology

Alice Yi-Szu Jou, Chen Liu, and Saeed Mohammadi

Purdue University, West Lafayette, IN, 47907, USA

Abstract — An integrated reconfigurable tuner based on a 4-mm long coplanar waveguide transmission line (CPW) loaded periodically with switched capacitors is demonstrated in a standard 45nm CMOS Silicon on insulator (SOI) technology. The tuner is designed with 10 switchable capacitors controlled by a series-in parallel-out shift register, that allows on-the-fly programming, leading to a large Smith chart coverage at 2 GHz with 1024 unique impedance points. The total area of the tuner is 4.4 ×0.42 mm^2 including DC and RF pads.

Index Terms — CMOS, Coplanar wave guide (CPW), Radio frequency, Reconfigurable, SOI, Tuner.

I. INTRODUCTION

Integrating multiple wireless communication standards in a single transceiver has long been a topic of high interest in radio frequency design. With continuously increasing wireless bands and modulation schemes, multiple switchable tank circuits are currently utilized to implement multimode multiband transceivers leading to high power consumptions and large chip areas. Reconfigurable RF band pass filters, low noise amplifiers (LNAs), power amplifiers, modulators and demodulators have been recently demonstrated [1-4]. Such reconfigurable modules enable future versatile reconfigurable transceivers.

Reconfigurable tuners have also been demonstrated [5-7] and not only find applications in reconfigurable transceivers, but can also be utilized as characterization modules for RF devices and circuits. As a variable impedance matching network at the input of a low noise amplifier, a reconfigurable tuner can provide optimum source impedance for minimum noise figure, maximum gain, or maximum linearity at multiple operating frequency bands. Moreover, a reconfigurable tuner can be placed at input (source-pull) and output (load-pull) of a power transistor/amplifier in order to achieve optimum power gain, efficiency and linearity as the operating frequency is varied [8]. Tuning mechanisms of reconfigurable tuners can be categorized into (i) magnetic tuning, (ii) mechanical tuning, and (iii) electronic tuning. While magnetic tuners such as yttrium-iron-garnet (YIG) and mechanical tuner such as evanescent mode cavity resonators present high quality factors (Q) and low loss, they suffer from large volumes and high power consumption, which make them unsuitable for integration with CMOS transceivers.

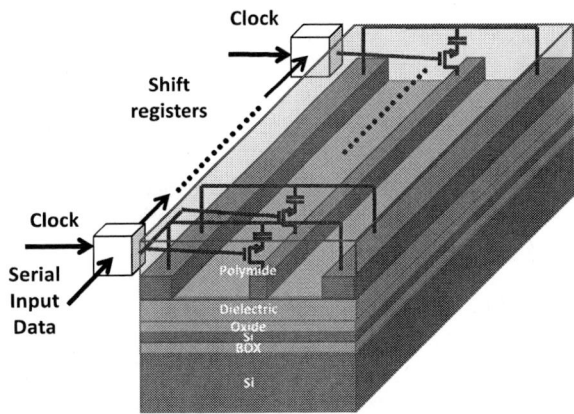

Fig. 1. Reconfigurable CMOS tuner with top metal layer to implement CPW transmission line and 10 segments of switched capacitor pairs and a 10-bit shift register control.

Electronic tuning using RF MEMS [5], GaAs varactors and barium-strontium-titanate (BST) varactors result in small size, high Q, high linearity and low loss. Nevertheless their integration with CMOS and BiCMOS transceiver circuits is still challenging. Reconfigurable tuners based on CMOS switches and CMOS varactors have been previously demonstrated in the frequency range of 4 to 16 GHz [6-7]. Designing reconfigurable tuners at lower frequencies (1-5 GHz) suitable for 2G to 4G wireless standards demand long and transmission lines. Such lines not only occupy a large area, but also suffer from excessive loss in CMOS technology.

Important factors in hindering the application of CMOS technology in reconfigurable RF and microwave circuits are (i) low quality factor Q and small capacitance tuning ratio of CMOS varactors, (ii) high parasitic capacitance of CMOS transistors and passive components due to conductive Si substrate, and (iii) lossy transmission lines due to high conductor losses of thin metallization and high dielectric losses of conductive Silicon substrate (resistivity of 1-10 Ω -cm). CMOS varactors can be avoided all together if series combinations of transistor switches and capacitors are utilized [7]. The parasitic capacitance of transistors and passive components can be minimized by utilizing a CMOS Silicon on Insulator (SOI) technology instead of a bulk CMOS technology. In designing a reconfigurable CMOS tuner, an optimization of various design parameters to address high losses of transmission lines is required to achieve a large Smith chart coverage.

978-1-4799-8198-4/15 $31.00 © 2015 IEEE

II. TUNER DESIGN

In [6-7], a CPW line is broken down into twenty two tunable sections using CMOS varactors or twenty sections using CMOS switch-capacitor pairs. The overall number of tunable impedance points on the Smith impedance chart are extremely high (2^{20} to 2^{22}), and far beyond the impedance points necessary for a practical tuner. The loss of so many varactors or CMOS switches in combination with their excess capacitive loading on the CPW line resulted in small coverage of Impedance Smith chart. In a practical load-pull tuner, any number of distinguishable impedance points on the Smith chart between 200 to 800 (which translates to 8 to 10 tunable elements) is enough to provide high enough density of impedances on the Smith Chart. Additionally, in [6] and [7], the tuners operate in the frequency ranges of 4-11 and 5-16 GHz, respectively. In this work, a reconfigurable tuner with large smith chart coverage suitable for wireless applications (1.7 to 2.6 GHz) is implemented in a standard 45nm CMOS SOI technology. Ten tunable section of a CPW line are designed using ten programmable switch-capacitor pairs to obtain 2^{10} distinguishable impedance points. A 10-bit serial-in parallel-out shift register is used to program the tuner in real time. The SOI technology reduces the parasitics of the transistor and passive components, resulting in improved on / off capacitance ratio, which leads to a better coverage of the Smith chart.

A. Tuning Mechanism

The tuner is composed of a 4 mm long CPW transmission line and 10 switchable capacitors (a transistor connected in series with a vertical natural capacitor). Each switch transistor is turned on/off by one of the flip-flops of the 10-bit shift register. By turning the series switch transistor on, a capacitor will be in series with a small resistor, leading to on capacitance of the small CPW segment $C_{on} = C_L$, where C_L is the series connected vertical natural capacitance. On the other hand, the load capacitance while transistor is in the off state equals to transistor drain to source capacitance C_{ds} in series with the vertical natural capacitance C_L:

$$C_{off} = \frac{C_{ds}C_L}{C_{ds}+C_L} \qquad (1)$$

C_{on} and C_{off} form on/off capacitance ratio of

$$\frac{C_{on}}{C_{off}} = 1 + \frac{C_L}{C_{ds}} \qquad (2)$$

In order to have wide Smith chart coverage, the on/off capacitance ratio should be large enough to separate various impedance points from each other. The low quality factor of the loading capacitor in the on-state Q_{Con} may also be detrimental in achieving large Smith chart coverage as low quality factors converge the impedance points towards each other inside a smaller area of the Smith chart. The quality factor is given by:

$$Q_{Con} = \frac{1}{\omega C_L R_{on}} \qquad (3)$$

where R_{on} is the transistor on resistance, inversely proportional to the transistor width and to drain-source capacitance C_{ds}.

A large on/off capacitance ratio (C_{on}/C_{off}) can be achieved by either large series capacitor C_L or small transistor width (small C_{ds}). Increasing C_L, however, would degrade the on-state capacitor quality factor (Eq. (3)) and leads to large parasitic capacitance to GND, which in turn, degrades the on/off capacitance ratio. Decreasing C_{ds} through small transistor widths benefits the on/off capacitance ratio but causes the on resistance of the transistor R_{on} to degrade leading to small on-state quality factor. As a result, there exists a trade-off between the on-state quality factor and on/off capacitance ratio, leading to very limited design space for such reconfigurable tuners.

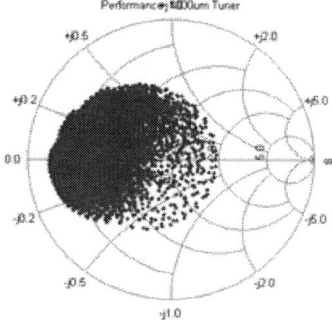

Fig. 2.   Matlab simulation with imported sNp file from Ansoft HFSS and Agilent ADS. Smith chart coverage of all 2^{10} impedance points at 2 GHz.

B. Switch capacitor design

The optimization of switch capacitor is accomplished by using a lumped model in Agilent ADS, and implemented with Cadence Virtuoso Spectre Circuit Simulator. For the targeted application in the frequency range of 1-5 GHz, the switch transistors were optimized to have a total width of $W_{total}/L = 805.6 \ \mu m/ 40$ nm with each transistor consists of 16 parallel connected 50 finger transistors (each finger width is 1 μm). Series connected capacitors $C_L = 2$ pF with compact dimensions of 34×24 μm^2 were designed using vertical natural capacitors available in this CMOS SOI technology.

C. Transmission line design

The 4-mm long transmission line is implemented with top two metal layers each with ~1.2 μm of Cu for low conductive loss. Above the top metal layers is a thin Polyimide passivation layer and beneath it is ~7 μm of dielectric material (SiO$_2$ and low-K dielectric polymer layers) and a 375 μm conductive Silicon substrate. The length of each segment of CPW line is 385 μm. The signal line width is optimized to W= 5 μm and the gap between the signal and ground lines is increased to G = 30 μm to achieve a high unloaded characteristic impedance of 80 Ω with finite ground plane width of S = 8 μm.

III. SIMULATION AND MEASUREMENT

A. Simulation

S-parameter data of transmission line stubs is first extracted from Ansoft HFSS 2-port S-parameter simulation files and then imported to Agilent ADS to simulate the transmission line stubs with lumped resistor and capacitor for on and off state switched capacitors. The final extracted S-parameter file of the whole circuit is then imported into Matlab to simulate Smith chart impedance coverage. The Smith chart coverage at 2 GHz is shown in Fig. 2, which includes 2^{10} impedance points. Post layout impedance points can be further simulated with Cadence Vituoso Spectre Simulator, which will appear closer to measurement.

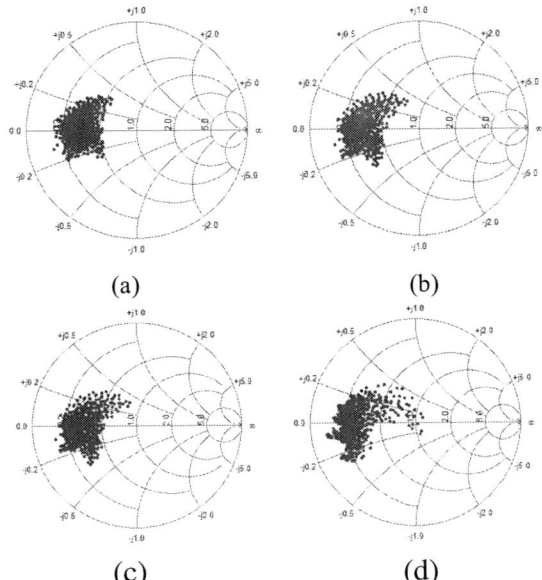

(a)   (b)

(c)   (d)

Fig. 3. 700 measured S_{11} points on the Smith chart using Agilent E8361A PNA at (a) 1.7, (b) 1.9, (c) 2 and (d) 2.6 GHz.

B. Measurement

The reconfigurable CMOS tuner is fabricated along a 10-bit serial-in parallel-out shift register in a standard 45 nm CMOS SOI technology with a total area of 1.85 mm^2. On wafer S-parameter measurements are done using an Agilent E8361A PNA Network Analyzer with short-open-load-thru (SOLT) calibration. Serial data and clock are generated by Arduino Uno board with ATmega328

microprocessor that feeds the shift register on the chip through its serial port. Fig. 3 shows about 700 measured impedance points at 1.7, 1.9, 2 and 2.6 GHz, where 1.7, 1.9 and 2.6 GHz frequencies are popular LTE bands in North America. The wide measured impedance coverage in the measured frequency bands shows a great opportunity to integrate the tuner directly to a reconfigurable LTE transceiver in a CMOS SOI platform.

IV. CONCLUSION

A reconfigurable tuner for the frequency range of 1-5 GHz fabricated in a standard 45 nm CMOS SOI technology is designed and its operation at frequencies around 2 GHz is demonstrated. The tuner structure uses a 4-mm CPW transmission line loaded with 10 transistor switches in series with capacitors. An integrated 10-bit serial-in parallel-out shift register enables programming the tuner to generate 2^{10} different impedance points. The transmission line and switch capacitors are optimized to have large Smith chart coverage. The small tuner dimension of 4 mm × 0.42mm allows its integration with reconfigurable transceivers.

REFERENCES

[1] A.Malczewski, B. Pillans, F. Morris, R. Newstrom, "A family of MEMS tunable filters for advanced RF applications," IEEE MTT-S Int. Microw. Symp. Dig., 2011 Jun., pp. 1–4.

[2] M. El-Nozahi,E. Sanchez-Sinencio, K. Entesari, "A CMOS Low-Noise Amplifier With Reconfigurable Input Matching Network," IEEE Trans. MTT, vol. 57 iss. 5, 2009 May, pp. 1054-1062.

[3] W.C.E. Neo, Y. Lin, X-D. Liu, L.C.N.de Vreede, L.E. Larson, M. Spirito, M.J. Pelk, K. Buisman, A. Akhnoukh, A. de Graauw, L.K. Nanver, "Adaptive Multi-Band Multi-Mode Power Amplifier Using Integrated Varactor-Based Tunable Matching Networks," IEEE JSSCC, vol. 41 iss. 9, 2006 Sept., pp. 2166-2176.

[4] V.K. Dao, Q.D. Bui, C.S. Park, "A Multi-band 900MHz/1.8GHz/5.2GHz LNA for Reconfigurable Radio," IEEE RFIC Symp., 2007 Jun., pp. 69-72.

[5] Q. Shen and S. Barker, "Distributed MEMS tunable matching networkusing minimal-contact RF-MEMS varactors," IEEE Trans. Microw. Theory Tech., vol. 54, no. 6, pp. 2646–2658, Jun. 2006.

[6] L. Rabieirad, S. Mohammadi, "A reconfigurable MEMS-less CMOS tuner for software defined radio," IEEE MTT-S Int. Microw. Symp. Dig., 2008 Jun., pp. 779-782.

[7] L. Rabieirad,S. Mohammadi, "Reconfigurable CMOS Tuners for Software-Defined Radio," IEEE Trans. MTT, vol. 57 iss. 11, 2009 Nov., pp. 2768–2774.

[8] F. Carrara, C. D. Presti, F. Pappalardo, g. Palmisano, "A 2.4GHz SOI CMOS Power Amplifier with Fully Integrated Reconfigurable Output Matching Network," IEEE Trans. MTT, vol. 57 No. 9, 2009 Sep., pp. 2122–2130.

Ferroelectric MIM Capacitors for Compact High Tunable Filters

Rosa De Paolis[1], Sandrine Payan[2], Mario Maglione[2], Guillaume Guegan[3], Fabio Coccetti[1]

[1] CNRS, LAAS, F-31400 Toulouse, France; Univ de Toulouse, LAAS, F-31400 Toulouse, France
[2] CNRS, Université de Bordeaux, ICMCB, F-33608 Pessac, France
[3] ST Microelectronics, 37100 Tours, France

Abstract — The voltage- and frequency-dependent material properties of (Ba,Sr)TiO₃ (BST) thin film have been extracted up to 67 GHz by means of a lumped elements equivalent circuit validated by fitting on de-embedded experimental data. The tunability is 73% (0-10 V bias), and the loss tangent is better than 0.06 (value at 0 V at 1 GHz). This varactor has been exploited in the design of a compact tunable filter implemented by using coupled resonators in lumped element in silicon planar technology. This solution yields a very compact size, and experimental results show a filter center frequency tuning of 83% (997 MHz – 1830 MHz) while maintain almost constant the fractional bandwidth at 35% upon the application of a 0-10 V bias. The insertion loss is between 3.8 and 4.4 dB, and the return loss is better than 14 dB.

Index Terms — ferroelectric materials; tunable circuits and devices; varactor; equivalent circuits; filters.

I. INTRODUCTION

Among the several solutions developed in the last years to address the increasing demand of frequency agility of modern wireless communications systems, ferroelectric materials, and in particular BST compounds, present advantages in terms of high power handling, continuous tuning, switching speeds, and operation in a large frequency range [1]. It is well known however that downsizing BST from bulk ceramic to thin film, affects permittivity, tunability, and dielectric losses. An accurate investigation of BST properties is therefore essential in view of more complex microwave circuits.

Reconfigurable and/or tunable filters are key elements in wireless communications systems and early works using BST material already exist [2]. One of the most promising design strategy allowing low insertion loss and high rejection is based on coupled resonators [3].

Extending the work done in one of these early designs [4], in this paper, a compact tunable filter has been fabricated by means of the sol-gel thin-film BST process developed at ST Microelectronics on 8-inches HRSi substrate. This technique is of high interest for industrial applications, due to the simple and cost-effective manufacturing process offered by the sol-gel approach.

A lumped element equivalent circuit applied to RF measurements on metal-insulator-metal (MIM) test capacitors, has been used to extract and validated the voltage-dependent material properties of the BST(permittivity, loss tangent, and tunability) up to 67 GHz, as described in Section II.

These MIM capacitors have been exploited in the implementation of the proposed tunable filter (Fig. 1) which experimental results are detailed in Section III. Compared to the current state-of-the-art on ferroelectric-based filters the proposed filter achieves excellent performance in terms of figure of merit (FoM), tunability, and total size.

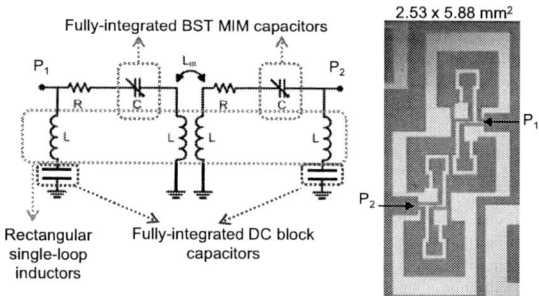

Fig. 1.   Equivalent circuit and layout (photo) of the proposed filter (total size 2.53 x 5.88 mm²).

II. EQUIVALENT CIRCUIT AND MATERIAL PROPERTIES EXTRACTION

A. Fabrication and Measurements Setup

The BST MIM capacitor is composed of a 200 nm-thick BST layer deposited by sol-gel process between two square 300 nm-thick platinum electrodes (Fig. 2). The CPW lines have been covered by a thicker metal layer of around 1.5 μm in order to reduce losses.

After a SOLT calibration, 2 ports S-parameters measurements have been performed by using the VNA (Anritsu 37397C) from 40 MHz to 67 GHz at room temperature, and under a continuous variable bias (0-10 V).

B. Equivalent Circuit and Parameters Extraction

As can be observed in Fig. 2, the capacitor is composed of the input/output pads (consisting of two symmetric fixed-length CPW lines), the core portion of CPW line containing the capacitor (actual DUT), and the interconnection between pads and MIM (represented by two asymmetric variable-length CPW lines). A cascade-based open-thru de-embedding method based on [5] has been developed and applied to measurement data in order to extract the intrinsic devices S-Parameters [6].

Finally, the resulting S-Parameters have been fitted with an RLC series equivalent circuit (EC) by means of two methods: a system of equations over frequency sweep by a MatLab routine, and a curve fitting optimization procedure in Agilent ADS simulator (semi-empirical approach). An excellent agreement has been achieved between measured and simulated data (both magnitude and phase) in the whole frequency band (from 40 MHz to 67 GHz) and over the applied voltages (0-10 V), thus validating the proposed approach up to 67 GHz (Fig.3).

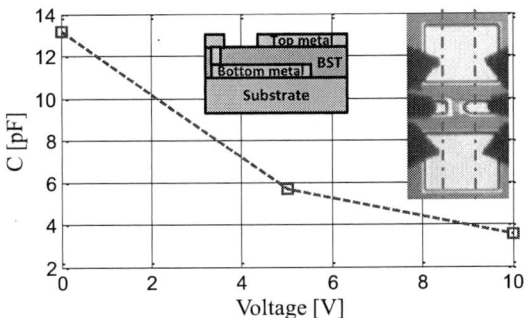

Fig. 2. Extracted capacitance versus voltage. A sketch and a photo of the MIM structure with reference planes for de-embedding are shown in the inset.

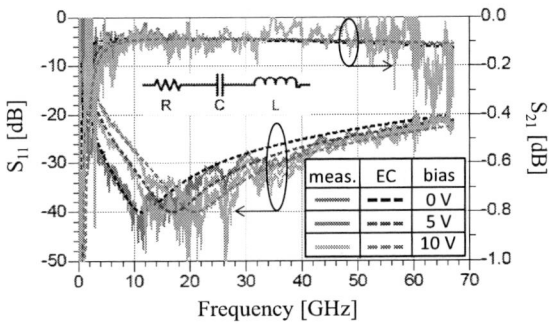

Fig. 3. Fitting in ADS of S_{11} and S_{21} parameters at {0,5,10} V: comparison between measurement data and simulations on the EC shown in the inset. Phases are not shown for brevity.

The extracted capacitance values clearly decreases upon the increasing of the bias voltage, as expected for a BST materials. Its value allows the estimation of the BST permittivity and tunability by considering the parallel plate configuration and by ignoring the fringing field (Fig.2).

The calculated permittivity is around 500 at 0 V; whereas the tunability (calculated as $(C(0)-C(V_{max}))/C(0)$) is around 73% between 0 and 10 V. As for the inductance and the resistance, they are, respectively, around 0.025 nH and 1 Ω over the entire bias range. Consequence of the lower losses at higher bias voltage the Q-factor of the de-embedded capacitors, extracted by the imaginary to real parts ratio of the de- embedded impedance improves as

expected (see Fig.4). The loss tangent, typically obtained as the inverse of the Q-factor, is below 0.06 at 1 GHz (value at 0 V). It is worth to notice that this latter includes the total losses of the device, of whereas the BST film loss is only a part of it.

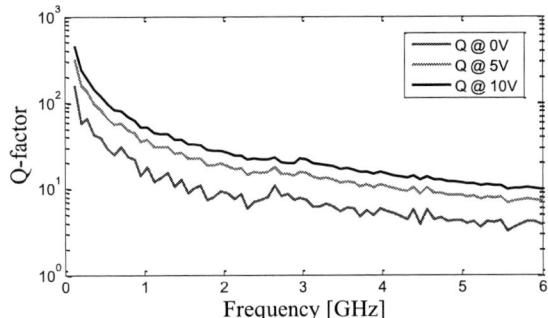

Fig. 4. Q-factor versus frequency at different bias voltage.

III. HIGH TUNABLE FILTER

The proposed tunable band-pass filter, is based on [4], and consists of two coupled lumped resonators. The equivalent circuit and the layout of the filter (total size 2.53 x 5.88 mm²) are presented in Fig. 1. The tuning element is the fully-integrated MIM capacitor presented in section II. On the other hand, the inter-resonators coupling is achieved by means of coupled inductors. More specifically, rectangular single-loop inductors have been chosen and their size, separation, and mutual orientation have been carefully optimized. Furthermore, decoupling capacitors have been integrated to allow bias application on BST capacitor while blocking undesired DC paths.

The tunable component enables the filter tuning: in particular, the center frequency f_0 of the filter (defined as the arithmetic mean of the two cutoff frequencies at 3 dB of attenuation) increases as the capacitance value decreases. The measured S-Parameters of the filter under the applied voltage are shown in Fig. 5.

Results demonstrate that, by applying a bias up to 10 V, the central frequency of the filter varies from 997 MHz to 1830 MHz, thus meaning a tunability (defined as $f_0(V_{min})$-$f_0(V_{max}))/f_0(V_{min})$) of 83%. Furthermore, under all the applied bias voltages an almost constant fractional bandwidth (defined as the ratio between the 3-dB bandwidth Δf and f_0) of 35% is obtained. The out of band rejection (at $2 f_0$ and at $3 f_0$) is always better than 20 dB. The insertion loss is between 4.4 and 3.8 dB, and the return loss is better than 14 dB.

Finally, a figure of merit for the overall characteristics of a tunable filter can be defined as in [2]:

$$FoM_{dB^{-1}} = \frac{f_0\left(V_D^{max}\right)-f_0\left(V_D^{min}\right)}{\sqrt{\Delta f\left(V_D^{min}\right)*\Delta f\left(V_D^{max}\right)}*\sqrt{IL\left(V_D^{min}\right)*IL\left(V_D^{max}\right)}}$$

TABLE I
STATE-OF-THE-ART OF FERROELECTRIC BANDPASS FILTERS

Ref.	f_0 [GHz]	Bias [V]	Frequency Tunability [%]	Bandwidth @ -3 dB [%]	Insertion loss [dB]	FoM [dB^{-1}]	FoM per bias [dB^{-1}/kV]
[2]	6.95	30	30.2	23.3-15.8	8.4-9.4	0.15	5.2
[7]	10.04	20	7.4	19.9-23.2	**2.9-1.95**	0.14	7
[8]	11.5	30	21.7	15.7-15.9	5.4-3.3	0.29	9.9
[9]	19.86	400	9.1	3.5-3.2	3.5-3.5	**0.73**	1.8
[10]	2.44	200	18	24.6-31.3	5.1-3.3	0.14	0.7
This work	1	**10**	**83**	35-34	4.4-3.8	0.41	**41**

where IL is the insertion loss at f_0. According to the previous expression, the FoM of the proposed filter is 0.41 dB^{-1}. Furthermore, if we relate this value to the DC bias applied, thus defining a FoM per applied kV of bias [2], it becomes 41 dB^{-1} kV^{-1}.

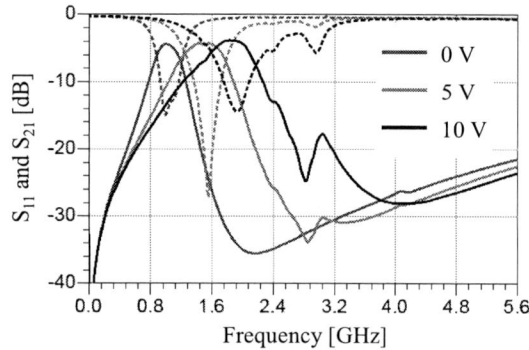

Fig. 5. Module of the S_{11} (dashed curves) and S_{21} (solid curves) of the filter at different value of bias voltage.

IV. CONCLUSION

In this paper BST thin films MIM capacitors have been characterized by means of a lumped elements equivalent circuit validated by fitting the de-embedded experimental data up to 67 GHz. At 1 GHz the tunability is 73% (0-10 V bias), and the loss tangent is better than 0.06 (at 0 V).

Based on these properties, a very compact high tunability filter has been proposed. To the best of authors' knowledge, the proposed filter achieve fair performance in terms of in-band insertion loss, but excels in tunability, FoM, and total size, if compared to the state-of-the-art of ferroelectric band-pass filters (reported in Table I).

ACKNOWLEDGEMENT

This work has been supported by the French National Agency (ANR) in the frame of the project ABSYS2 (n°ANR 2010 VERS 012). The authors also thank Mitsubishi Materials Corporation for sol-gel solutions supply.

REFERENCES

[1] S. Gevorgian, Ferroelectrics in Microwave Devices, Circuits and Systems, London: Springer, 2009.

[2] S.L. Delprat, J.H. Oh, F. Xu, L. Li, E.E. Djoumessi, M. Ismail, M. Chaker, and K.Wu, "Fully Distributed Tunable Bandpass Filter Based on Ba$_{0.5}$Sr$_{0.5}$TiO$_3$ Thin-Film Slow-Wave Structure," International Journal of Microwave Science and Technology, 2011.

[3] R. Stefanini, M. Chatras, P. Blondy, G.M. Rebeiz, "Compact 2-pole and 4-pole 2.4–2.8 GHz dual-mode tunable filters," IEEE MTT-S International Microwave Symposium Dig., pp. 1480-1483, May 2010.

[4] R. De Paolis, F. Coccetti, S. Payan, A. Rousseau, M. Maglione, G. Guegan, "Microwave Characterization of Ferroelectric Thin Films for Novel Compact Tunable BST Filters," IEEE European Microwave Conference (EuMC), Oct. 2013.

[5] M-H Cho, G-W Huang, K-M Chen, A-S Peng, "A novel cascade-based de-embedding method for on-wafer microwave characterization and automatic measurement," IEEE MTT-S Int. Microwave Symp. Dig., vol. 2, June 2004.

[6] R. De Paolis, F. Coccetti, S. Payan, M. Maglione, G. Guegan, "Characterization of Ferroelectric BST MIM Capacitors up to 65 GHz for a Compact Phase Shifter at 60 GHz," IEEE European Microwave Conference (EuMC), Oct. 2014.

[7] S. Courreges, Y. Li, Z. Zhao, K. Choi, A. T. Hunt, and J. Papapolymerou, "Two-pole X-band-tunable ferroelectric filters with tunable center frequency, fractional bandwidth, and return loss," IEEE Trans. on Microwave Theory and Techniques, vol. 57, no. 12, pp. 2872–2881, 2009.

[8] J. Papapolymerou, C. Lugo, Z. Zhiyong, X. Wang, and A. Hunt, "A miniature low-loss slow-wave tunable ferroelectric BandPass filter from 11–14 GHz," IEEE MTT-S Intern. Microw. Symp. Dig., San Francisco, June 2006.

[9] V. N. Keis, A. B. Kozyrev, M. L. Khazov, J. Sok, and J. S. Lee, "20 GHz tunable filter based on ferroelectric (Ba,Sr)TiO$_3$ film varactors," Electronics Letters, vol. 34, no. 11, pp. 1107–1109, 1998.

[10] J. Nath, et al., "An electronically tunable microstrip bandpass filter using thin-film Barium Strontium Titatnate (BST) varactors," IEEE Transactions on Microwave Theory and Techniques, Vol. 53, no. 9, pp. 2707-2712, 2005.

10.6 THz Figure-of-Merit Phase-change RF Switches with Embedded Micro-heater

Jeong-Sun Moon, Hwa-Chang Seo, Dustin Le, Helen Fung, Adele Schmitz, Thomas Oh, Samuel Kim, Kyung-Ah Son, and Baohua Yang

HRL Laboratories, Malibu, CA, 90265, USA

Abstract—We report on GeTe-based phase-change RF switches with embedded micro-heater for thermal switching. With heater parasitics reduced, GeTe RF switches show on-state resistance of 0.05 ohm*mm and off-state capacitance of 0.3 pF/mm. The RF switch figure-of-merit is estimated to be 10.6 THz, which is about 15 times better than state-of-the-art silicon-on-insulator switches. With on-state resistance of 1 ohm and off-state capacitance of 15 fF, RF insertion loss was measured at <0.2 dB, and isolation was >25 dB at 20 GHz, respectively. RF power handling was >5.6 W for both on- and off-state of GeTe.

Index Terms—RF switches, phase-change material, wireless communications, power handling, insertion loss.

I. INTRODUCTION

RF switches are ubiquitous in RF systems. Key features of RF switches include low insertion loss, high isolation, excellent linearity, power handling, easy integration with conventional semiconductor technologies, and reliability and packaging. Currently, RF switches include SOI [1], SOS [2], GaN [3] FET switches and RF MEMS [4]. For comparison, typical RF switch $R_{on} \cdot C_{off}$ values are 230-300 femtosecond for RF SOI switches, 448 for RF SOS switches, 453 for GaN FET switches, and ~4 for RF MEMS switches. RF MEMS switches offer the best FOM with excellent linearity >70 dBm. For high RF power handling, GaN FET switches showed 40 W continuous-wave RF power handling with <0.3 dB of compression [3].

Emerging GeTe-based phase-change materials (PCM) are being evaluated for implementation in RF switches [5]. GeTe RF switches are distinguished by resistance change between their amorphous (high resistance) and crystalline (low resistance) phases. The static resistance ratio is on the order of ~10^6. With GeTe switches fabricated in vertical via configurations, RF insertion loss of 0.66 dB was reported at 10 GHz with its third-order intercept point (IIP3) of 37 dBm [6]. Several GeTe RF switches have been reported with excellent RF switch FOM, $1/(2\pi * R_{on} * C_{off})$, of >100 THz [7] as an intrinsic switch FOM. On the other hand, with a micro-heater embedded for thermal actuation, the switch FOM varies due to parasitics, ranging from 1 THz [8], >4 THz [9], and 7.3 THz [10]. RF power handling of these GeTe RF switches both at on-state and off-state has not been fully investigated. So far, the maximum on-state RF power handling ranges from >0.6 W [8], >20 dBm [9], and >2 W [7, 10-11]. Off-state RF power handling has not been reported yet. Also GeTe switch RF stability under continuous-wave RF soak is not known.

In this talk, we report on GeTe-based RF switches on silicon substrates with an embedded micro-heater with reduced parasitics, yielding FOM of 10.6 THz—the highest to date—and continuous-wave RF power handling of >5.6 W for GeTe on-state and of 10 W for GeTe off-state. Under 0.5 W continuous RF soak, almost no drift in insertion loss was observed. While PCM-based RF switches are at a very early development stage, our report shows that GeTe PCM RF switches are promising for future RF front-ends, offering excellent integration with any RF substrates.

II. PCM RF SWITCHES

The GeTe PCM material is deposited on SiO_2/Si wafers. As deposited, the GeTe is amorphous with sheet

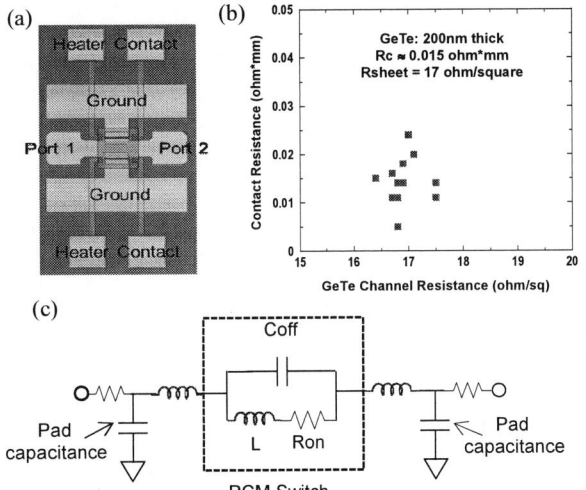

Figure 1. (a) An optical photograph of fabricated GeTe RF switch in a shunt configuration, (b) Measured contract resistance versus GeTe channel resistance, (c) A network of resistance of the GeTe RF switch

resistance >1 MΩ/sq. Figure 1(a) shows a fabricated GeTe phase-change RF switch in a shunt configuration with embedded micro-heater. With GeTe material patterned by ICP dry etching for isolation, Ti-based ohmic contact electrodes are formed. The ohmic contact resistivity is measured by the standard transmission-line-method (TLM), which yields a record low lateral contact resistance of ~15 Ω·μm, as shown in Figure 1(b). Figure 1(b) also shows measured GeTe sheet resistance of 17

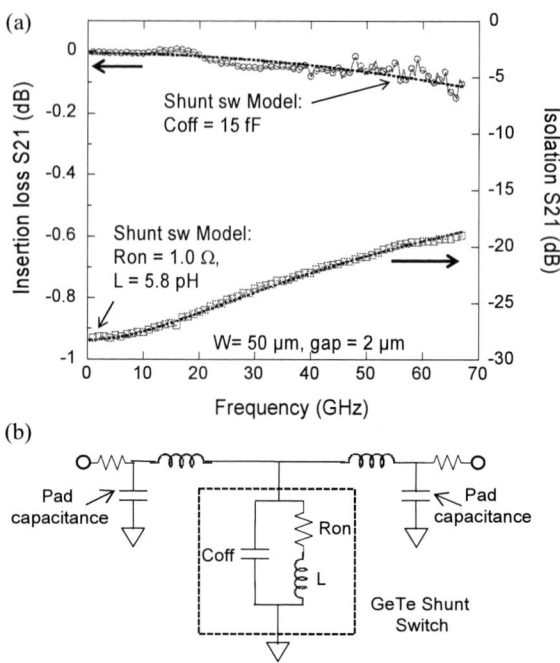

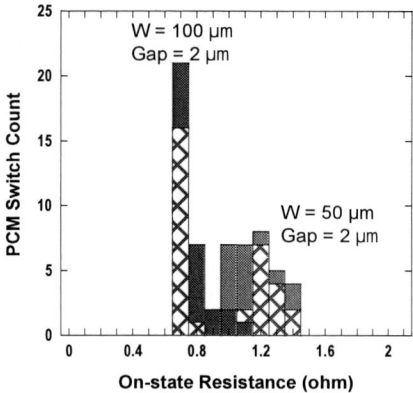

Figure 2. Measured on-state resistance of GeTe phase chance RF switches

ohm/sq after annealing above 200°C. The on-state resistivity of GeTe material used here is 3.4 x 10^{-4} Ω·cm. Figure 1(c) shows a lumped element circuit to explain the on-state resistance, consisting of contact resistance and GeTe channel resistance. Figure 2 shows measured on-state resistance of GeTe RF switches with 4-terminal DC resistance measurement. The on-state resistance values are <~0.7 Ω for 0.1 mm channel width, and ~1.2 Ω for 0.05 mm channel width. With total channel width of 0.1 mm, the R_{on} is ~0.07 Ω·mm, which is the lowest among FET-based RF switches.

III. MEASUREMENT RESULTS

A. RF Characteristics and Modeling

Figure 3(a) shows measured and modeled S-parameters of GeTe RF switches up to 67 GHz for both GeTe with on-state and off-state. The channel width was 50 μm with a gap of 2 μm. The measurements were done with a standard SOLT calibration on CS-5 impedance standard. For GeTe on-state (shunt sw off), the isolation was 26 dB at 20 GHz. The S21 is modeled with R_{on} and inductor L, as shown in Figure 3(b). The extracted GeTe on-state resistance was 1.0 Ω, which is very close to DC values. With total channel width of 0.05 mm, R_{on} was 0.05 Ω·mm. The inductor L was 5.8 pH. For GeTe off-state (shunt sw

Figure 3. (a) Measured and modeled s-parameters of GeTe RF shunt switches with switch-on and switch-off states, (b) An equivalent circuit with lumped elements of Ron, Coff, and inductor L, used to model the s-parameter data

on), the insertion loss was 0.16 dB at 20 GHz. The S21 was modeled with C_{off} of 15 fF (0.3 pF/mm), as shown in Figure 3(b). Figure 3(b) shows an equivalent circuit model with lumped elements of R_{on}, L, and C_{off} shown in Figure 3(a). The parasitic pad capacitance and inductors was extracted with standard 'open' and 'short' devices on the same wafer. The measured and modeled data (in Black) are in excellent agreement up to 67 GHz.

With R_{on} of 1.0 ohm and C_{off} of 15 fF, the RF switch FOM is 10.6 THz, showing the potential to enable RF switch to millimeter wave frequencies, as shown in Figure 3(a).

B. RF power handling and Stability

Figure 4 plots measured output RF power versus input RF power for GeTe RF switches at 2 GHz with both GeTe off-state and GeTe on-state. When the GeTe is off-state, a 0.05-mm GeTe RF switch can handle input RF power up to 10 W, showing a 1 dB compression point (P_{1dB}) of >10 W. With GeTe PCM RF switch width of 0.05 mm, the off-state P_{1dB} is greater than 80 W/mm. Figure 4(b) shows RF power handling with GeTe on-state. A 0.05-mm GeTe RF switch can handle input RF power up to 2.8 W, and a 0.1-

mm GeTe RF switch shows input RF power handling >5.6 W.

For the first time, GeTe RF switches have been characterized in terms of RF stability under constant continuous-wave input power of 0.5 W. Figure 5 shows a plot of insertion loss over time for one week, showing the insertion loss was not changed at all over one week of 0.5 W RF soak.

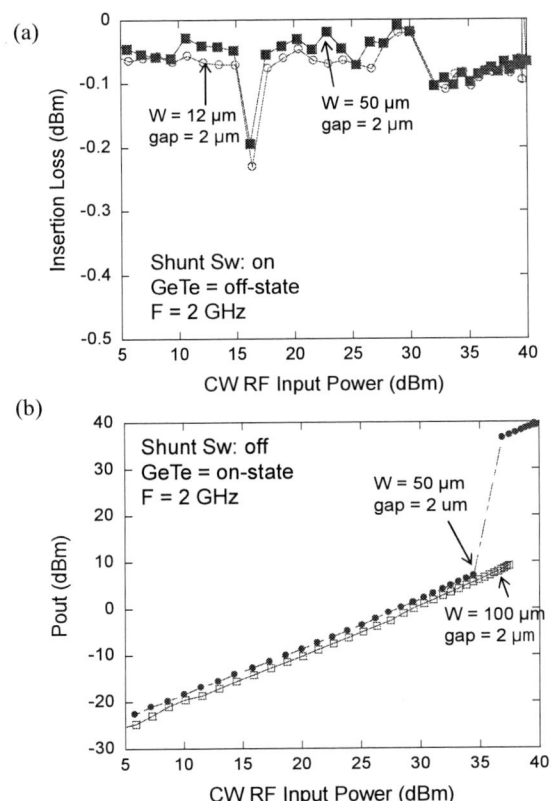

Figure 4. Measured continuous-wave RF power handling of GeTe PCM RF switches at (a) GeTe off-state and (b) GeTe on-state

VI. CONCLUSION

For the first time, we fabricated a new kind of RF switch using GeTe phase-change material with very low on-state resistance of 0.06 Ω·mm. The insertion loss, linearity and RF power handling dramatically improved over the previously reported PCM-based vertical via switches. The overall RF performance is state-of-the-art, and makes the PCM RF switches highly promising.

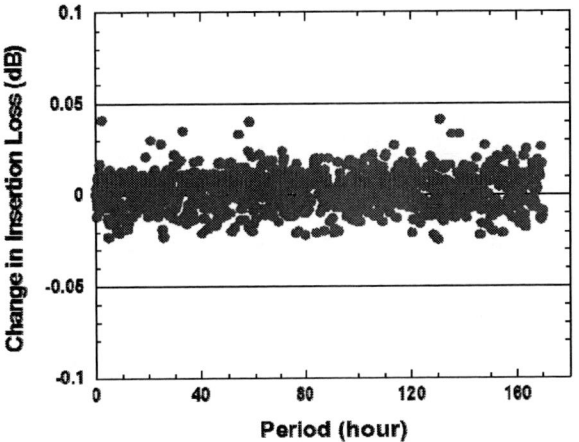

Figure 5. Measured change in the insertion loss of GeTe RF shunt switch under 0.5 W RF power for one week with GeTe off-state

REFERENCES

[1] A. Boutla et al., "A Thin-film SOI 180 nm CMOS RF Switch Technology", 2009.

[2] D. Kelly, C. Brindle, C. Kemerling, and M. Stuber, "The State-of-the-art of Silicon-on-Sapphire CMOS RF switches", CSICS Digest, pp. 200 - 203, 2005.

[3] C. F. Campbell and D. C. Dumka, "Wideband High power GaN on SiC SPDT Switch MMIC", IEEE MTT-S 2010 International Microwave Symposium Digest, pp. 145-148, 2010.

[4] G. Rebeiz et al., "Tuning in to RF MEMS" IEEE Microwave Magazine, pp. 55-72, 2009.

[5] H. Lo et al., "Three-Terminal Probe Reconfigurable Phase-change Material Switches", IEEE Trans. Electron Dev., vol 57, pp. 312 – 320, 2010.

[6] Y. Shim, G. Hummel, and M. Rais-Zadel, "RF switches using phase change materials", IEEE International Conference on MEMS Digest, pp. 237-240, 2013.

[7] J. S. Moon et al., "Development toward high-power sub-1 ohm DC-67 GHz RF switches using phase change materials for reconfigurable RF front-end", IEEE MTT-S 2014 International Microwave Symposium Digest, pp. 1-3, 2014.

[8] N. El-Hinnawy et al., "A four-terminal, inline, Chalcogenide Phase-change RF Switch using an independent resistive heater for thermal actuation", IEEE Electron. Dev. Lett., vol. 34, p. 1313, 2013.

[9] M. Wang and M. Rais-Zadeh, "Directly heated four-terminal phase-change switches", IEEE MTT-S 2014 International Microwave Symposium Digest, pp. 1-4, 2014.

[10] N. El-Hinnawy et al., "A 7.3 THz Cut-off frequency, inline, Chalcogenide Phase-change RF switch using an independent resistive heater for thermal actuation", Compound Semiconductor IC Symposium (CSICS), pp. 1-4, 2013.

[11] J. S. Moon et al., "High-linearity 1-ohm RF switches with phase-change materials", IEEE Silicon Monolithic Integrated Circuits in RF Systems (SiRF), pp. 7-9, 2014.

978-1-4799-8198-4/15 $31.00 © 2015 IEEE

Low power and High speed OOK Modulator for Wireless Inter-Chip Communications

Hae Jin Lee, Chong Hyun Yoon, Joong Geun Lee, Chae Jun Lee, Dong Min Kang, In Sang Song, Sung Jun Cho, Hong Yi Kim, Inn Yeol Oh, and Chul Soon Park.

Department of Electrical Engineering, KAIST, 291 Daehak-ro, Yuseong-gu, Daejeon 305-701, Korea

Abstract — Millimeter-wave on-off keying (OOK) modulator for wireless inter-chip communications is proposed. The proposed modulator not only has a wide bandwidth which allows the high speed modulation but also operates with a low power consumption enhancing the power efficiency and assures the inter-chip communication. With wide bandwidth and the OOK modulation scheme, it can process data having 12 Gbps while consuming a little DC power showing the energy efficiency of 0.75 pJ/bit. The on/off ratio of the circuit is higher than 25 dB assuring little degradation on the SNR. The chip size using 65nm GP RF CMOS is 0.254 mm² which is very compact compared to other state-of-the-art works.

Index Terms — CMOS, millimeter-wave. Chip-to-chip interconnect, OOK modulation, high data rate.

I. INTRODUCTION

Recently, the amount of data that needs to be processed among electronic chips is increasing sharply. The data rate capability of inter-chip data communication is shown in Fig. 1. Current inter-chip data communication is achieved via electrical wires. The electrically wired data communication has met a barrier of high data rate processing. A single electrical wire can only process a certain amount of data at a given time. To process data having its speed exceeding the limited data rate that a single wire can handle, the only solution is to place the wires in parallel. However, it may cause the cross talk between parallel connected wires.

There have been many attempts to replace the current electrically wired inter-chip communication. One of them is the inductive coupling [1]. The inductive coupling has a great prospect of being the alternative thanks to the low power consumption and processing data with high data rate but has a disadvantage of very short communication distance that is less than hundreds of microns. The other choice for the next inter-chip communication is to utilize the optical interconnection [1]. It shows a great advantage in achieving a large data rate while having little signal loss throughout the optical cable. However, the optical interconnect has additional power consumption and is difficult to integrate a silicon substrate with optical-to-electrical converter and vice versa.

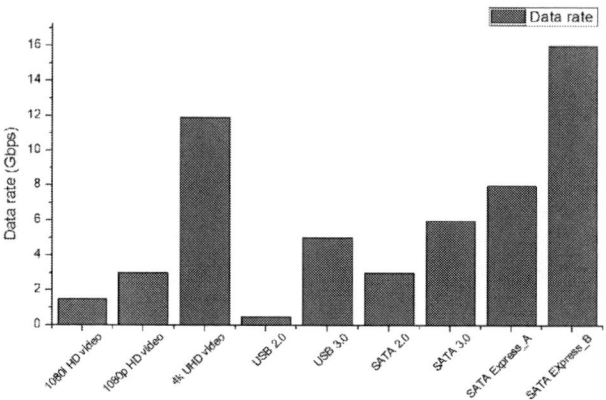

Fig. 1.   Different processing-capable data rates of different inter-chip data interface

The optimum alternative for the current inter-chip communication using electrical wires is the wireless interconnection. The wireless interconnection can overcome the disadvantages of current wired interconnection and many attempts stated above.

Wireless interconnect using millimeter-wave (mm-Wave) above 60 GHz has received attention for inter-chip communication because of its wideband characteristic. Using these frequency bands, high data rate wireless communication can be feasible.

The most important performance of inter-chip communication system is energy efficiency. If its power consumption is very large in spite of high-data capability, the system cannot be applicable for inter-chip communication. To lower the energy efficiency, OOK modulation and sub-nm CMOS technology are adopted for the low power operation system.

II. CIRCUIT DESIGN

Typical OOK modulator shown in Fig. 2 is the switching amplifier structure which is suitable for the inter-chip interconnect system thanks to the possibility of processing a high data rate and high dc power efficiency. However, there are drawbacks that need to be taken into account. First, low on/off ratio must be taken care of because LO input signal can be leaked to RF output through M2 even if M2 is off. This degrades on/off ratio and then results in lowering signal-to-noise ratio. Second, conventional structure shows low energy efficiency that cannot be applicable for inter-chip communication. Therefore, to achieve high speed and low power consumption, high on/off ratio and low energy efficiency must be satisfied.

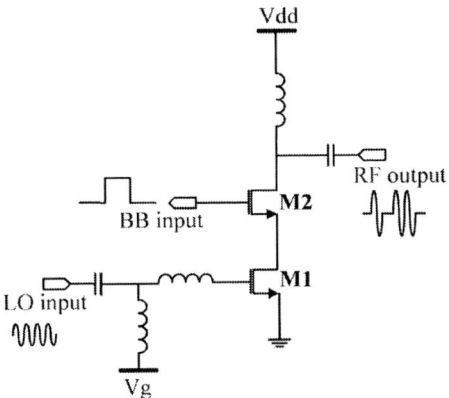

Fig. 2. Conventional mm-Wave OOK modulator; switching amplifier structure

Using 65 nm GP RF CMOS technology, the proposed mm-Wave OOK modulator for low power consumption and high data rate was designed.

Fig. 3 shows the proposed mm-Wave OOK modulator circuit. To overcome low power efficiency and low on/off ratio of conventional structure, current reuse technique for low power consumption and PMOS switching scheme for high on/off ratio are adopted.

Current reuse means that two-stage common source amplifier M1 and M2 shares a same dc-current path. When baseband input turn on M2 transistor, LO signal is amplified through M1, M2 transistors and appears to RF output node. When M2 is turned off, LO signal cannot pass through M2 so that no output signal appears at the output node. Although M2 is turned off, a little amount of the amplified LO signal leaks through the turned off transistor M2 and will be detected at the output node resulting in lowering on/off ratio. To reduce this leakage signal, a PMOS transistor M3 is inserted at the output node whose drain is connected to RF output node, and source is connected to VDD. When baseband input logic is low, M2 turns off, and the applied PMOS turns on and keeps the RF output node to AC ground sharply resulting in ensuring the output to be kept to zero.

Through the inverter, a baseband signal is increasing from VDD to ground so that and M2 transistor can switch quickly. Also the size of M2 transistor is designed to achieve a high data rate of baseband signal.

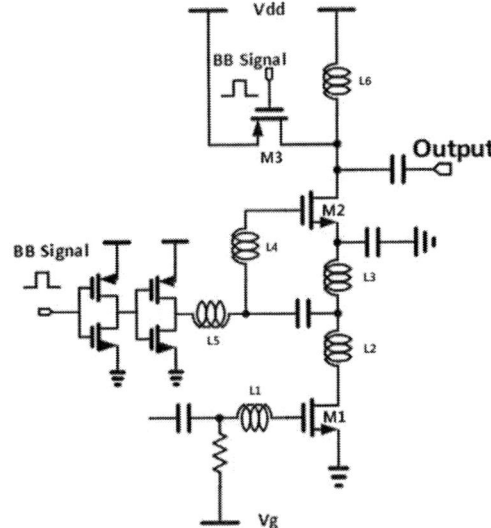

Fig. 3. The proposed mm-Wave OOK modulator; current reused modulator with high isolation

All of inductors are used for impedance matching, and micro-strip lines are used as the transmission line. MIM capacitors are used as the AC ground capacitor.

Simulation results show that maximum modulated data rate is 12Gbps and output 1dB compression power is -3.76dBm at 77GHz shown in Fig. 4.

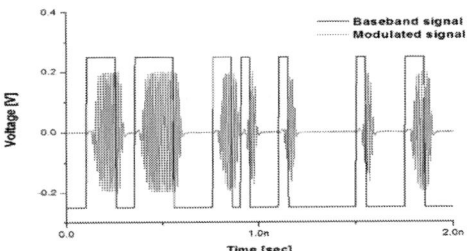

(a) Transient simulation response

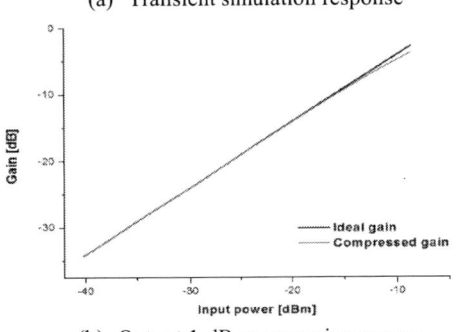

(b) Output 1-dB compression power

Fig. 4. Simulation results of the proposed OOK modulator; (a) transient response (b) Output 1-dB compression power

978-1-4799-8198-4/15 $31.00 © 2015 IEEE

TABLE I
COMPARISON WITH THE STATE-OF-THE-ART WORKS

Reference	Technology	Frequency [GHz]	Modulation	Data rate [Gbps]	P_{DC} [mW]	EE [pJ/bit]	Supply [V]	Chip size [mm²]
[2]	40nm CMOS	57,80	ASK	20	52	2.6	1.1	0.025
[3]	40nm CMOS	56	OOK	11	14.7	0.83	1.1	0.15
[4]	90nm CMOS	60	OOK	3.5	5	1.43	-	-
[5]	90nm CMOS	60	OOK	3.3	100	30.3	1.2	-
[6]	90nm CMOS	60	OOK	2	14.4	7.2	1.8	0.58
This work	**65nm CMOS**	**77**	**OOK**	**12**	**9**	**0.75**	**1.0**	**0.254**

III. MEASUREMENT RESULTS

Fig. 5 shows a chip photograph, and the chip size is 0.254 mm². The measured S-parameter results is shown in Fig. 6 and was measured by using Anritsu VNA using a frequency extender. Maximum conversion gain (S_{21}) is 5 dB and 3-dB bandwidth is 28 GHz. return loss S_{11} is 5 dB from 65 to 110GHz. DC power consumption is 9mW so that the energy efficiency under 12Gbps is 0.75pJ/bit. Mismatches between simulation and measurement results comes from the mismatch between passive device modeling and measurement results.

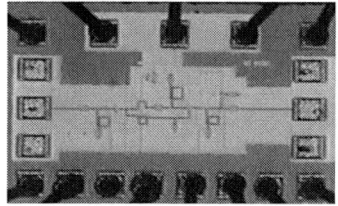

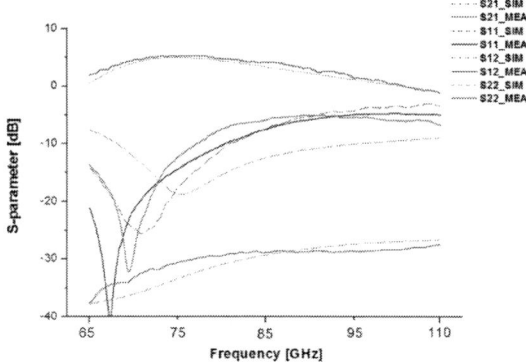

Fig. 5. Chip photograph of the proposed modulator

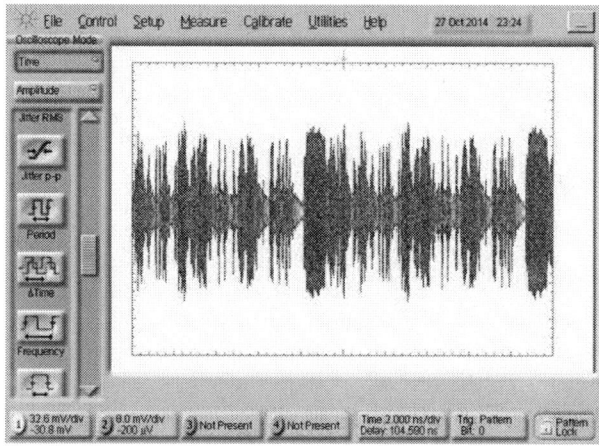

Fig. 6. Measured S-parameter results compared with simulation results

The measured modulation signal is shown in Fig. 7 and was measured by using Agilent 86100C Oscilloscope. Because modulated output signal is directly connected to the oscilloscope, a modulated signal envelope can be confirm through this measurement. The figure shows the modulator stably modulates for 12Gbps signal.

Fig. 7. Measured the modulated signal for 12-Gbps baseband data.

Table I compares the measured results of the proposed modulator with that of the state-of-the arts, and the proposed OOK modulator shows lowest energy efficiency under 12Gbps data rate.

IV. CONCLUSION

A low power consuming and high data rate processing optimized millimeter-wave OOK modulator for inter-chip communication application has been presented. The proposed modulator adopts the OOK scheme and the switching amplifier structure with current reused two-stage common source amplifier to lower the power consumption. Also a PMOS has been presented at the output node to enhance the performance of the modulator both at "ON" and "OFF" state. The proposed OOK modulator would be suitable for the future wireless inter-chip data communication which arises as an alternative for the current wired inter-chip communication.

978-1-4799-8198-4/15 $31.00 © 2015 IEEE

ACKNOWLEDGEMENT

This research was funded by the MSIP (Ministry of Science, ICT & Future Planning), Korea in the ICT R&D program 2014.

REFERENCES

[1] C. Byeon, "Millimeter-wave low-power transceiver designs for short-range communications," *Ph. D Dissertation, dept. Electrical engineering, KAIST*, 2013.

[2] Y. Tanaka, et al., "A versatile multi-modality serial link," *IEEE International Solid-State Circuits Conference, Dig. Tech. Papers*, pp. 332-334, Feb. 2012.

[3] K. Kawasaki, et al., "A millimeter-wave intra-connect solution," *IEEE Journal of Solid-State Circuits*, vol.45, no.12, pp.2655-2666, Dec. 2010.

[4] E. Juntunen, et al., "A 60-GHz 38-pJ/bit 3.5-Gb/s 90-nm CMOS OOK digital radio," *IEEE Transactions on Microwave Theory and Techniques*, vol.58, no.2, pp.348-355, Feb. 2010.

[5] J. LEE, et al., "A low-power low-cost fully-integrated 60-GHz transceiver system with OOK modulation and on-board antenna assembly," *IEEE Journal of Solid-State Circuits*, vol.45, no.2, pp.264-275, Feb. 2010.

[6] J. LEE, et al., "Gbps 60GHz CMOS OOK modulator and demodulator," *IEEE Compound Semiconductor Integrated Circuit Symposium*, Oct. 2010.

A 20GHz Class-C VCO Using Noise Sensitivity Mitigation Technique

Kento Kimura, Kenichi Okada and Akira Matsuzawa

Department of Physical Electronics, Tokyo Institute of Technology
2-12-1-S3-27, Ookayama, Meguro-ku, Tokyo 152-8552 Japan.
Tel: +81-3-5734-3764, Fax: +81-3-5734-3764
E-mail: kimura@ssc.pe.titech.ac.jp

Abstract— **This paper presents a Class-C VCO with noise sensitivity mitigation technique. Class-C VCO has a large parasitic capacitances between gate and source nodes and this capacitance variation causes a large frequency sensitivity to noise voltage. As a consequence, the noise from gate node become the largest noise contributor. Proposed technique can control this sensitivity by tuning the tail impedance. A 65nm CMOS prototype of the VCO demonstrates oscillation frequency from 19.35 to 22.36 GHz, the phase noise of -105.8 dBc/Hz at 1MHz offset with power dissipation of 8.7mW and Figure-of-Merit of -182.4 dBc/Hz.**

I. INTRODUCTION

Oscillators are key components for most of the modern electrical devices. Their output noise which randomly appears in phase is a main performance limiter of a communication systems. Reducing power dissipation has become a crucial design requirement for mobile devices that are not constantly connected to a power source. For both achievement, Class-C VCO [1] in Fig.1(a) is the promising in terms of the effective impulse sensitivity function (ISF). Further Class-C VCO can allocate almost all power to the fundamental frequency component and other harmonic can be negligible. However the necessity to keep the transistors in saturation region for perfect Class-C operation limits the swing magnitude and also causes the start-up difficulty. One breakthrough of this issue is the additional adaptive biasing circuits [2], [3] in Fig.1(b) to operate after oscillation. These circuits initially set high gate voltage for robust start-up and shift to low gate voltage for power reduction after oscillation started. However phase noise performance deteriorates with decreasing the gate voltage because of a quite large cross coupled transistors to provide large trans-conductance during small conduction time.

This paper presents a phase noise degradation mechanism of Class-C VCO and proposes new technique to improve phase noise performance. The detail of relevancy between noise sensitivity and transistor size are investigated in Section II. In next Section, the details of proposed technique and the measurement results are presented.

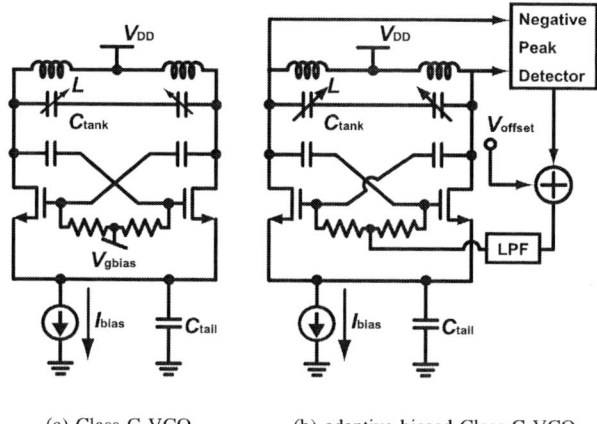

(a) Class-C VCO (b) adaptive biased Class-C VCO

Fig. 1. The circuit schematics of Class-C VCOs

II. INVESTIGATION OF PHASE NOISE DEGRATATION

Several phase noise theories have already been reported and one of a degradation mechanisms is AM-PM conversion on varactor. VCO gain (K_{VCO}) amplifies noises from the charge pump and this effect becomes so crucial with large K_{VCO}. Equation (1) shows quantification of the phase noise degradation by AM-PM conversion.

$$\mathcal{L}(\omega_{offset}, K_{VCO}) = 10 \log_{10} \left(\frac{K_{VCO} V_m}{2 \omega_{offset}} \right)^2 \quad (1)$$

where V_m is the noise voltage and ω_{offset} is the offset frequency.

A same phenomenon can be observed in gate bias nodes in Class-C VCO. The shift of gate-bias voltage causes frequency shift because parasitic capacitances of the cross-coupled pairs also shift. Here this frequency shift ratio is defined as K_{VGBIAS}.

$$K_{VGBIAS} = \frac{\partial \omega}{\partial V} = -\frac{\omega_0}{2C} \frac{\partial C_{CCTr}}{\partial V} \quad (2)$$

where ω_0 is the oscillation frequency and C_{CCTr} is the capacitance of the cross-coupled pairs.

The cross coupled transistors have 4 large parasitic capacitances C_{GS}, C_{GB}, C_{GD} and C_{DB} and we neglect C_{DS} and C_{SB} which are quite smaller than 4 others. This is because a tail transistor of Class-C VCO should be large enough to operate in linear region and the drain voltage of a tail transistors is almost similar to ground potential. Assuming the back gate of transistors is connected to ground, C_{SB} can be negligible. C_{GS}, C_{GB} and C_{DB} are connected serially and only C_{GD} is connected parallel as shown in Fig. 2.

$$C_{CCTr} = \frac{1}{2}C_{GS} + \frac{1}{2}C_{GB} + \frac{1}{2}C_{DB} + 2C_{GD} \qquad (3)$$

In Class-C VCO, DC-cut capacitors C_{DC} should be taken into consideration. However C_{DC} are large enough to guarantee large amplitude and we can suppose $C_{DC} \gg C_{GS} > C_{GD}, C_{GB}, C_{DB}$. As a consequence, Equation (3) can also be applied in Class-C VCO. Among these capacitances, the only C_{GS} largely change between off-region and saturation region as shown in Fig. 3. Equation (4) shows the only C_{GS} dominantly affect the capacitance slope of C_{CCTr}.

$$\frac{\partial C_{CCTr}}{\partial V_{GBIAS}} = \frac{1}{2}\frac{\partial C_{GS}}{\partial V_{GBIAS}} \qquad (4)$$

Further C_{GS} is expressed as equation (5). The capacitance gap between saturation and off region become larger with increasing size of transistor. Driving with lower gate voltage requires larger width transistor for robust oscillation and a noise sensitivity to the gate noise largely

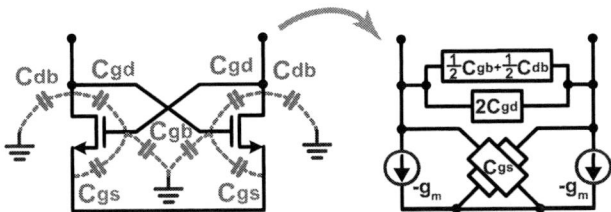

Fig. 2. The contribution of parasitic capacitance

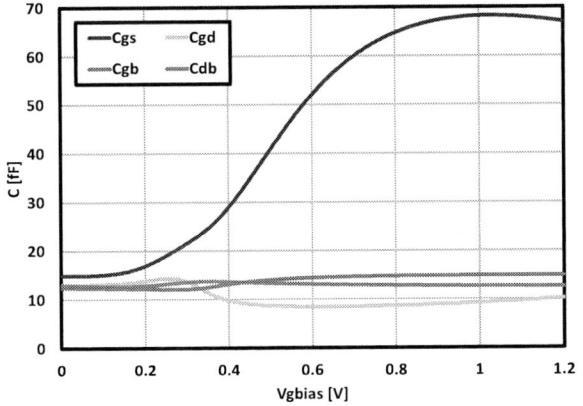

Fig. 3. The voltage dependences of each parasitic capacitances

increase in this case.

$$C_{GS} = \begin{cases} \frac{2}{3}WLC_{OX} & (V_{GS} > V_{TH}) \\ WL_{OV}C_{OX} & (V_{GS} < V_{TH}) \end{cases} \qquad (5)$$

where W, L is the width and length of the transistors, C_{OX} is the oxidative capacitance, L_{OV} is the length of diffusion overlap and V_{TH} is the threshold voltage.

III. Proposed Sensitivity Mitigation Technique

As we discussed in the previous section, shift of the C_{GS} has to be minimized. Fig.4(a) shows the proposed circuit schematic which has the additional impedance between the drain nodes of tail transistors. The effect of C_{GS} to total capacitance changes according to the Z because only C_{GS} is connected to the drain node of the tail transistors. Further C_{TAIL} cannot be negligible in proposed circuits because each drain nodes are not directly connected. Equation (6) shows the coefficient of C_{GS} will be replaced and C_{GS} slope can be controlled by tuning a tail impedance.

$$C'_{GS} = \left\{ 1 - \frac{g_m Z + \omega^2 C_{GS}(C_{GS} + C_{TAIL})Z^2}{1 + \omega^2 (C_{GS} + C_{TAIL})^2 Z^2} \right\} C_{GS} \qquad (6)$$

where Z is the tail impedance, g_m is the transconductance of the cross coupled pairs and C_{TAIL} is the capacitance connected to the drain of tail transistors.

This impedance block is composed of 4 bit switch resistors and can be tuned from 5 ohm to 1000 ohm. The coefficient can largely change when Z is between 10 and 100. Even if only C_{GS} slope become flat, the total slope cannot be flat because the characteristics of other parasitic capacitances become dominant in this case. We have to cancel each parasitic capacitance characteristics by tuning the coefficient of C_{GS}. Fig.5 shows the characteristics of the cross-coupled pairs capacitance in proposed circuits. When Z is 60, the total noise sensitivity can be almost zero. Fig.6 shows the measurement results of this circuits. Compared to conventional Class-C VCO, the proposed method improves 3 dBc/Hz on 1MHz offset frequency.

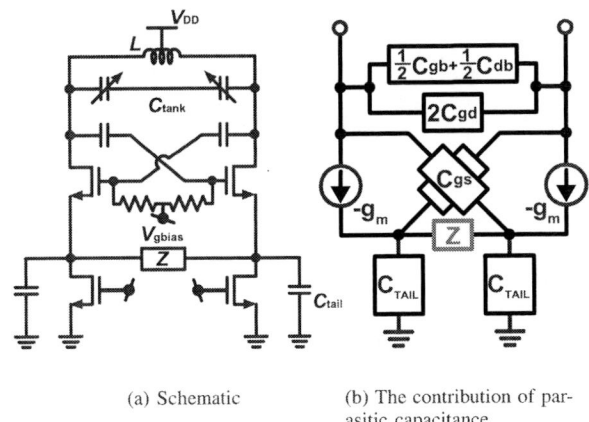

(a) Schematic

(b) The contribution of parasitic capacitance

Fig. 4. Proposed Class-C VCO

978-1-4799-8198-4/15 $31.00 © 2015 IEEE

TABLE I
PERFORMANCE SUMMARY AND COMPARISON WITH RECENTLY REPORTED 20 GHz LC OSCILLATORS.

Ref.	PhaseNoise[dBc/Hz]	Frequency [GHz]	Power [mW]	FoM [dBc/Hz]	Technology	Topology
[4]	-101@1MHz	26.7	21	-176.3	65nm CMOS	push-push
[5]	-98@1MHz	18.7	6	-176	65nm CMOS	PMOS
[6]	-112@1MHz	19	200	-174.5	0.13um BiCMOS	Colpitts
[7]	-106@1MHz	17.9 - 21.2	19.2	-179	65nm CMOS	Tail Capacitive Feedback
This work w/o NSM	-102.4@1MHz	19.3 - 22.4	8.9	-179	65nm CMOS	Class-C
This work w. NSM	-105.8@1MHz	19.3 - 22.4	8.7	-182.4	65nm CMOS	Class-C with NSM

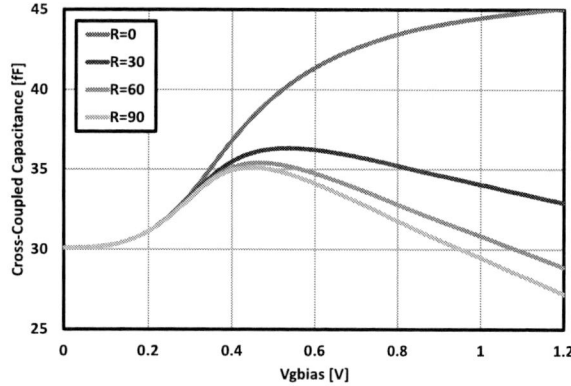

Fig. 5. The capacitance characteristics of the cross coupled pairs in proposed circuits

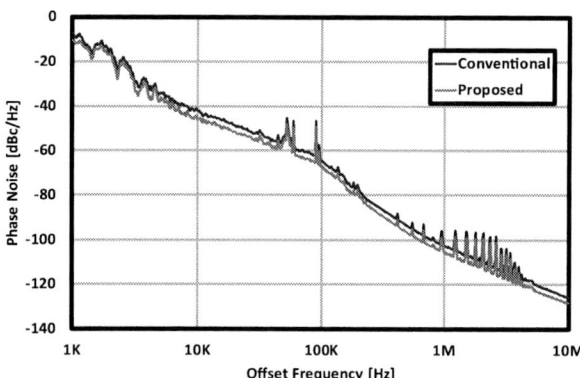

Fig. 6. Phase Noise Comparison

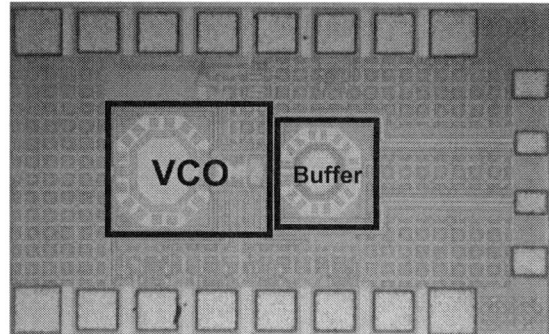

Fig. 7. Chip Photograph

IV. CONCLUSION

The mechanism of phase noise degradation in class-C VCO with low gate voltage is a high frequency sensitivity to gate noises. Especially the shift characteristics of C_{GS} is quite sharp around V_{TH} and capacitance of the cross-coupled pairs also have similar characteristics. The proposed approach which tunes the tail impedance can control this frequency sensitivity and the noise effects from the bias resistors and adaptive biasing circuits can be mitigated. The phase noise performance is improved 3dBc/Hz while consuming only 8.7mW with the area occupation 0.057 mm², and the best Figure of Merit in 20GHz band is achieved.

ACKNOWLEDGMENT

This work is partially supported by MIC, SCOPE, MEXT, STARC, STAR and VDEC in collaboration with Cadence Design Systems, Inc., and Mentor Graphics, Inc.

REFERENCES

[1] A. Mazzanti and P. Andreani, "Class-C harmonic CMOS VCOs, with a general result on phase noise," IEEE Jounal of Solid-State Circuits, vol.43, no.12, pp.2716–2729, Dec 2008.

[2] W. Deng, K. Okada, and A. Matsuzawa, "A feedback class-C VCO with robust startup condition over PVT variations and enhanced oscillation swing," Europoan Solid State Circuits Conference, pp.499–502, Sep 2011.

[3] W. Deng, K. Okada, and A. Matsuzawa, "Class-C VCO With Amplitude Feedback Loop for Robust Start-Up and Enhanced Oscillation Swing," IEEE Journal of Solid-State Circuits, vol.48, no.2, Feb 2013.

[4] R. Molave, S. Mirabbasi, and H. Djahanashahi, "A 27-GHz Low-Power Push-Push LC VCO with Wide Tuning Range in 65nm CMOS," IEEE Int. Symp. Circuits and Systems, pp.1141–1144, May 2011.

[5] G. Zhu, S. Diao, F. Lin, and D. Guidotti, "A Low-Power Wide-Band 20GHz VCO in 65nm CMOS," 5th Global Symposium on Milimeter Waves, pp.291–294, May 2012.

[6] W. Wang, Y. Takeda, T. Teh, and B. Floyd, "A 20GHz VCO and Frequency Doubler for W-band FMCW Rader Applications," IEEE Silicon Monolithic Integrated Circuits in RF Systems, pp.104–106, Jan 2014.

[7] A. Musa, R. Murakami, T. Sato, W. Chaivipas, K. Okada, and A. Matsuzawa, "A Low Phase Noise Quadrature Injection Locked Frequency Synthesizer for MM-Wave Applications," IEEE Journal of Solid-State Circuits, vol.46, no.11, pp.2635–2649, Nov 2011.

Radio-Frequency Flexible Transistors on Cellulose Nanofibrillated Fiber (CNF) Substrates

Jung-Hun Seo[1], Tzu-Hsuan Chang[1], Ronald Sabo[2], Zhiyong Cai[2], Shaoqin Gong[3], Zhenqiang Ma[1]

[1]Department of Electrical and Computer Engineering, [2]USDA Forest Products Laboratory, Madison, WI 53726, [3]Wisconsin Institute for Discovery, University of Wisconsin-Madison
Email: mazq@engr.wisc.edu, zcai@fs.fed.us, sgong@engr.wisc.edu

Abstract — RF performance flexible thin-film transistors toward green portable devices were realized. The cellulose nanofibrillated fiber (CNF) substrate combined with Si nanomembranes (Si NMs) printing technique enables to fabricate flexible, high-speed and bio-degradable devices. Flexible Si NM thin-film transistors (TFTs) built on the CNF substrate show mobility of 336 cm/v·s and f_T and f_{max} of 2.4 GHz and 5.1 GHz, respectively. This demonstration paves the path to entire green portable devices so as to generate less waste and save more valuable resources.

Index Terms — Cellulose nanofibrillated fiber, Si nanomembranes, bio-degradable and flexible device.

I. INTRODUCTION

The use time of portable gadgets gets shorter as new technology has continuously developed and more stylish gadgets are made. Users tend to change their cellphones in every 2-3 years and a large quantity of working portable electronics is discarded. As an example, more than 426,000 cellphones were discarded per day in U.S [1]. A large amount of natural resource is being consumed while polluting our environment. The majority of the components in portable gadgets is built on Si wafer which is highly purified and expensive substrate. In fact, only a tiny portion of the Si material is used in electronics.

Commercial standard 12 inch Si wafer has a thickness of 775 μm but, only top 1 μm is used for electronic devices. The rest of the expensive substrate is just for support of devices and is eventually wasted when devices are discarded. Therefore, it is economical to make electronics on the self-decomposed platform, rather than on Si.

In this paper, we report Si nanomembranes (Si NMs) based flexible RF transistor built on the Cellulose Nanofibrillated Fiber (CNF) substrate which is made from wood. Because CNF substrate was made from cellulose of wood and can be formed as a thin-sheet, it has good bendability and bio-degradability. It is a suitable substrate for entire green portable electronics [2, 3]. To realize RF performance thin-film transistors (TFTs) on the CNF substrate, active materials with both high mobility and flexibility are required. Si NMs have few hundreds of nm thick and can be printed onto flexible substrate to fabricate high performance flexible TFTs [4,5]. One of the advantages of Si NMs is that it inherits the properties of bulk Si while maintaining the flexibility and durability [6,7]. Overall Si NMs is one of the best materials to realize RF performance bio-degradable devices.

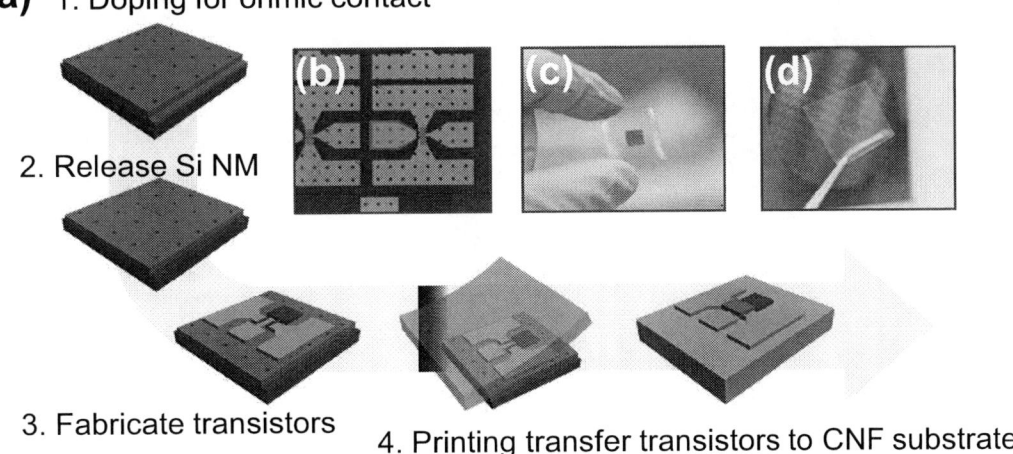

Figure 1. Illustration of fabrication process for flexible SiNM TFTs built on the CNF substrate. (b) Microscope image of finished device on the CNF substrate, (c-d) optical images show the transparency.

II. EXPERIMENTAL

As shown in Fig. 1, fabrication starts with phosphorus ion implantation into SOI wafer with its top Si being lightly p-type doped. Implanted samples were diffused and recrystallized in the furnace and formed n+ well for ohmic contact of source/drain metal. Top Si layer was released from buried oxide layer by selective wet-etching using concentrated HF. Released top Si layer, called Si NM from this step, gently fell down on the handling Si substrate. After rinsing sample, it was dried by N_2 blowing. A 0.8 μm length gate electrodes and SiO dielectric stacks were patterned by e-beam lithography and deposited by e-beam evaporator, followed by deposition of source/drain electrode. After the completion of metallization step, entire devices were transfer printed onto the adhesive layer coated CNF substrate and cured to finish the device fabrication as shown in Fig. 1 (b)-(d). Fig. 1 (c)-(d) shows transparency of substrate. Detailed processes about transfer printing and TFT fabrication can be found in elsewhere [8].

III. RESULT AND DISCUSSION

Fig. 2(a) shows the microscopic image of finished device on a CNF substrate. Double gate fingers were used. A physical gate length and effective gate (channel) length is 0.8 μm and 0.5 μm, respectively. The gate width is 20 μm, Since all metallization processes were finished before transfer, Si NMs was faced up while all electrodes were

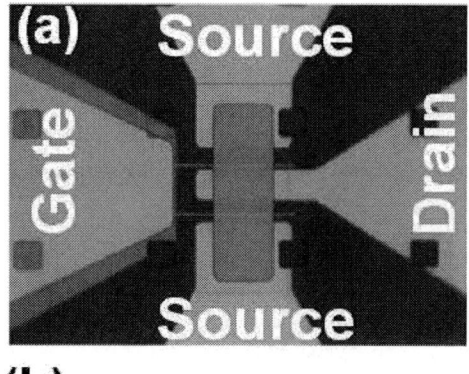

Figure 2. (a) Microscopic image of finished flexible Si NM TFTs on the CNF substrate. (b) Cross section of device. Due to the transfer printing, Si NMs were faced up.

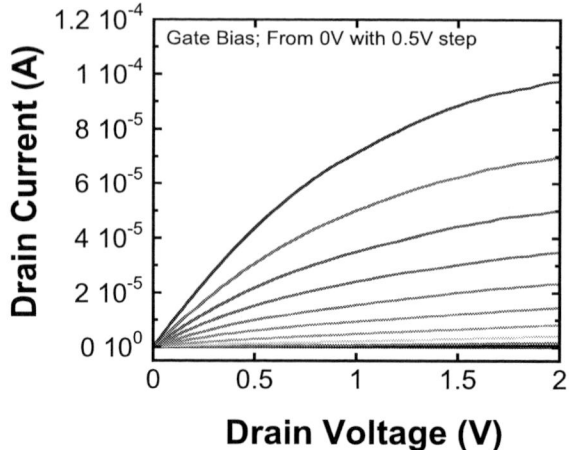

Fig. 3. Output characteristics under different Vg ranging from 0 V to 5 V with a 0.5 V step.

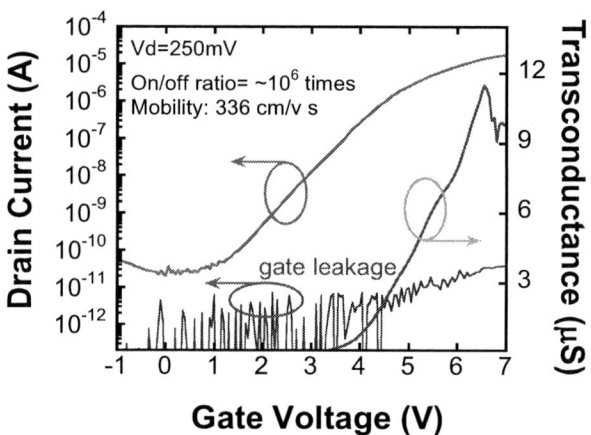

Fig. 4. Transfer curves, gate leakage current and transconductance (g_m) curves at a drain bias of 250 mV.

faced down as shown in Fig. 2 (b). Fig. 3(a) shows the output curve from TFTs under various gate biases from 0V to 5V. Good saturation of output curve shows that drain current was well controlled by gate bias. Fig. 3(b) shows the transfer curve, gate leakage current and transconductance (g_m) versus gate voltage (V_g). The highest g_m is 13 μS under drain bias of 250 mV. 10^6 time of on/off ratio indicates very low off-current was realized. 336 cm/V·s of field-effect mobility was measured based on the following equation:

$$\mu = \frac{L_g \cdot g_m}{W_g \cdot C_{ox} \cdot V_{ds}}$$

Gate leakage current remains very low within all gate bias range.

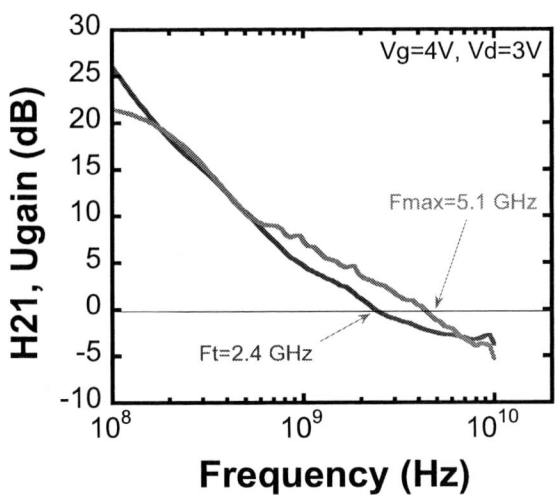

Fig.5 Measured frequency response characteristics of Si NM TFTs

Fig. 5 shows the measured frequency response characteristics of Si NM TFTs under V_{gs}=4V and V_{ds}=3V. The f_T and f_{max} are 2.4 and 5.1 GHz, respectively. Small C_{gs} and C_{gd}, due to the small overlap between gate and source/drain, can be achieved by the precise alignment during e-beam lithography and are attributed to high RF characteristics, The performance of the device is currently limited by the large critical features of the devices. Upon reduction of the device dimensions, higher performance can be readily expected.

IV. CONCLUSION

In conclusion, high performance Si NM flexible built on the CNF substrate is demonstrated. Devices show good DC and RF characteristics and the results suggest wood based green substrates can be used as high performance transistor supporters. The concept implies future portable devices/gadgets can be made biodegradable and also can be applicable to a wide range of portable devices.

ACKNOWLEDGEMENT

The work is supported by AFOSR under grant FA9550-09-1-0482. The program manager is Dr. Gernot Pomrenke.

REFERENCES

[1] "The Big Picture: 426 000 Discarded Cellphones," *IEEE Spectrum*, vol.44, no.9, 24-25, Sept. 2007.
[2] J.Y. Zhu, R. Sabo, X.L. Luo, "Integrated Production of Nano-fibrillated Cellulose and Biofuel (Ethanol) by Enzymatic Fractionation of Wood Fibers", Green Chemistry 13(5) 1339-1344, 2011.
[3] R. Sabo , J.-H. Seo, Z. Ma, "Cellulose nanofiber composite substrates for flexible electronics" 2012 TAPPI International Conference on Nanotechnology for Renewable Materials. Montreal, Quebec 2012.
[4] L. Sun, G. Qin, J.-H. Seo, G. K. Celler, W. Zhou, and Z. Ma, "12-GHz Thin-Film Transistors on Transferrable Silicon Nanomembranes for High-Performance Flexible Electronics." *Small*, 6, 2553–2557, 2010.
[5] H. Zhou, J.-H. Seo, D. M. Paskiewicz, Y. Zhu, G. K. Celler, P. M. Voyles, W. Zhou, M. G. Lagally, and Z. Ma. "Fast flexible electronics with strained silicon nanomembranes." *Scientific reports*, 3 1291, 2013.
[6] Y. Sun, W. M. Choi, H. Jiang, Y. Y. Huang, and J. A. Rogers, "Controlled buckling of semiconductor nanoribbons for stretchable electronics." *Nature Nanotechnology*, 1, 201–207, 2006.
[7] Z. T. Zhu, E. Menard, K. Hurley, R. G. Nuzzo, and J. A. Rogers, "Spin on dopants for high-performance single-crystal silicon transistors on flexible plastic substrates." *Applied Physics Letters*, 86, 133507, 2005.
[8] K. Zhang, J.-H. Seo, W. Zhou, and Z. Ma. "Fast flexible electronics using transferrable silicon nanomembranes." *Journal of Physics D: Applied Physics*, 45, 14 143001, 2012.

Phase Noise Reduction in RF Oscillators Utilizing Self-Injection Locked and Phase Locked Loop

Li Zhang, Ajay K. Poddar*, Ulrich L. Rohde*, and Afshin S. Daryoush

Dept. of ECE, Drexel University, Philadelphia, PA 19104 USA
*Synergy Microwave Corp., 201 McLean Boulevard, Paterson, NJ 07504 USA
POC: daryoush@coe.drexel.edu

Abstract — Self-injection locked and phase locked loop (SILPLL) using long optical delay has been proposed for phase noise reduction in microwave oscillators. Experiments for both single loop and dual loop SILPLL are performed to demonstrate the phase noise reduction in an electrical VCO. Dual loop configuration provided a phase noise reduction of 71 dB at 300 Hz offset while achieving a lower spurious signal level as opposed to single loop configuration.

Index Terms — forced oscillation, self-injection locked and phase locked loop, phase noise, VCO, fiber optic link,

I. INTRODUCTION

Conventional forced oscillation techniques of injection locking (IL) and phase-locked loop (PLL) are two viable methods for further phase noise reduction by introducing external frequency reference to the oscillator being stabilized [1]-[2]. While IL technique are easy to implement, the phase noise in the close-in offset frequency range is degraded due to frequency detuning and limited locking range as explained in [3]. On the other hand, even though PLL has a longer pull-in time that results in a slower response, the high gain loop filter enables the PLL to remove the close-in to carrier phase noise significantly, while far away from carrier offset suffers from a higher noise. Sturzbecher et al. demonstrated that in externally forced oscillators, a better phase noise characteristics for both close-in and far-away offset frequencies and a wider locking range are achieved by combining IL and PLL (ILPLL) [4]. However, external reference sources are required in the conventional ILPLL topology, which limits the ultimate phase noise performance.

To bypass the limit of external source, self-forced oscillation techniques have been proposed and developed. In [5], a self-injection locking (SIL) topology is proposed by passing the output of oscillator through an electrical delay line and feeding it back using a circulator. The experimentally verified modeling demonstrates that the overall oscillator phase noise is inversely proportional to the signal delay time. However, due to the high loss in electric delay lines, the phase noise improvement is not significant. To avoid lossy electrical delays, we have recently reported self-injection locking (SIL) employing long optical fiber delay lines [6]-[7] by taking advantage

of the high Q factor from long length and low loss optical fiber. Our approach provides substantial phase noise reduction as opposed to SIL employing electrical delay lines. The concept of forced oscillation could also be extended to self-phase locked loop (SPLL) demonstrated by Pillet et al. [8]. In the proposed structure, the microwave signal generated from the beat note of a dual frequency laser (DFL) was sent into a delay line frequency discriminator (DLFD) whose output is used to stabilize the laser frequency for generating a more stable beat note. The authors have also demonstrated that SPLL technique can also be applied to OEOs and electrical VCOs to provide significant phase noise reduction [9].

In this paper, a novel technique to achieve a cleaner spectrum in a wider offset frequency range has been proposed by combing simultaneously SIL and SPLL, i.e. self-injection locked and phase locked loop (SILPLL). This paper documents experimental results of SILPLL that the close-in to carrier phase noise of SILPLL is reduced over that of SIL alone by adding the SPLL operation, while far-away phase noise is reduced due to the SIL. The concept of SILPLL is extended to a dual loop SILPLL VCO employing a delay combination of 3km and 5km [10]-[11].

II. SILPLL SYSTEM OVERVIEW

An electrical VCO is built to investigate the phase noise performance of SILPLL technique. A tunable band pass filter (BPF) is manufactured on a FR4 substrate (shown in Fig. 1) to provide VCO frequency tuning function which is achieved by changing the bias voltage of the varactor diode. The VCO is realized by connecting the BPF to the input and output of an amplifier (Avantek AMT 9634). The VCO oscillates at 8.6GHz with an output power of 0dBm, and the tuning sensitivity of this VCO is 5.5 MHz/V at 1 V and 1 MHz/V at 5 V. Measured phase noise of this free-running VCO is shown in the dashed curve of Fig. 4. The phase noise has a slope of about 30 dB/decade, indicating a flicker noise dominated system. The spot noise is -11 dBc/Hz at 300 Hz offset and -58 dBc/Hz at 10 kHz offset.

978-1-4799-8198-4/15 $31.00 © 2015 IEEE

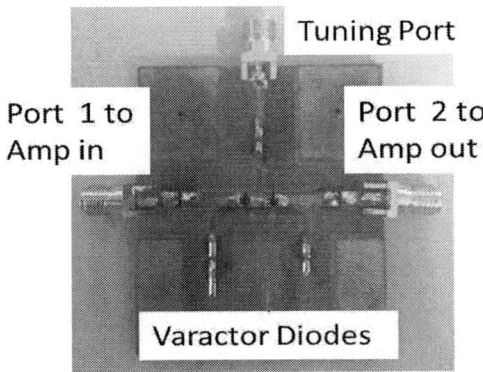

Fig. 1. Image of the tunable BPF. Tuning is achieved by changing the varactor diode reverse bias voltage at the tuning port

The optical delay in SILPLL system is provided by a fiber optic link using electro absorption modulator (EAM), as shown in Fig. 2 with blue blocks. A laser diode is integrated with an EAM in a single package (Fujitsu FLD5F10NP), denoted as 'EAM-LD' in the block diagram. Output of the 'EAM-LD' is amplified by an EDFA (NuPhoton NP2000) whose output is split into two paths with different delays. Two photodetectors (PD) terminates the optical delay to perform the optical to electrical conversion. The EAM link has a loss of 12 dB and a NF of 40 dB.

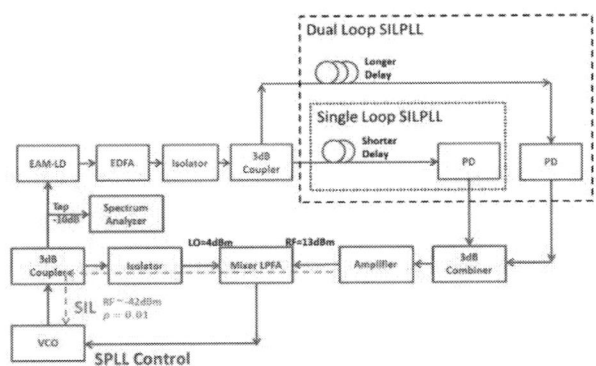

Fig. 2. Block diagram of VCO with SILPLL using EAM fiber optic link.

For the SILPLL operation, the EAM-LD is driven by the VCO, and the modulated light passes through either a single optical delay line, termed as single loop SILPLL or two different optical delay lines, termed as dual loop SILPLL (DSILPLL). The delayed signal is first amplified by an amplifier (Avantek AMT 9634) and is compared against the non-delayed VCO output to generate an error signal for frequency adjustment, completing the SPLL function (purple path in Fig. 2). The error signal is generated by the 'Mixer + LPFA' board that contains a double-balanced mixer as phase detector and an op-amp

circuit to serve as low pass filter amplifier (LPFA). A 'Mixer+LPFA' board with a low pass cutoff frequency of 20kHz is used for the SILPLL experiment. More detail of the 'Mixer+LPFA' board can be found in [8]. Power level at mixer RF port is about 13 dBm, resulting in a phase detector sensitivity of 0.1 V/rad. Note that in Fig. 2, there is not a dedicated path for SIL function. However, the finite isolation of 35 dB from mixer RF port to LO port and an isolation of 20 dB from the isolator result in an injection power level of -42 dBm to the VCO, corresponding to injection strength $\rho = 0.003$. Since the BPF has a low Q of about 100, the VCO is prone to injection locking even for small injection strength. Image of the experimental setup is provided in Fig. 3.

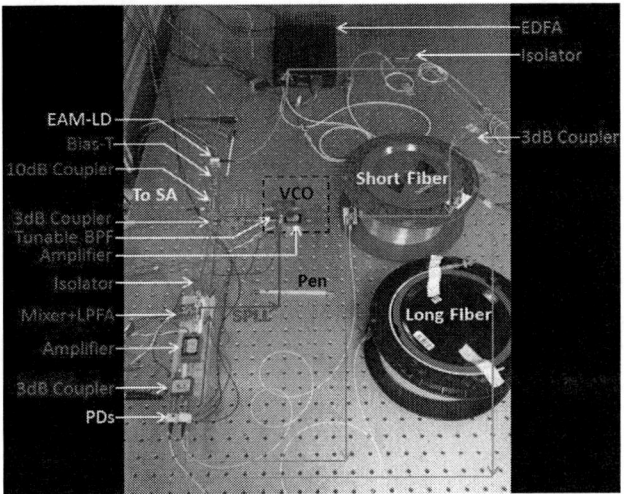

Fig. 3. Block diagram of VCO with SILPLL using EAM fiber optic link.

III. EXPERIMENT WITH SILPLL VCO

Measured phase noise of single loop SILPLL with 1 km delay is plotted as blue curve in Fig. 4. The achieved phase noise is -69 dBc/Hz at 300 Hz corresponding to a reduction of 58 dB and is -87 dBc/Hz at 10 kHz corresponding to a reduction of 29 dB. Note that when the SPLL is functioning, the SIL due to power leakage from mixer RF port is also present; the overall phase noise is really due to the combination of SIL and SPLL. For comparison, phase noise of SIL alone with 1km delay is also provided in red curve of Fig. 4. We can see that SILPLL phase noise is superior to SIL phase noise up to 3 kHz offset, beyond 3 kHz SILPLL phase noise follows SIL phase noise which is expected from the analytical prediction [4]. Fiber delays longer than 1 km are also attempted to provide more phase noise reduction. However, the side-modes of the long delay makes the PLL loop unstable. To reduce the side-mode level of long delay for further phase noise reduction, dual loop SILPLL is illustrated next.

Phase noise performance of DSILPLL with 3 km and 5 km delays is shown in green curve in Fig. 4. Phase noise of -82 dBc/Hz at 300 Hz offset is achieved, resulting in a 71 dB improvement; at 10 kHz offset, the phase noise is -98 dBc/Hz, resulting in a 40 dB improvement. As we can see from Fig. 4, DSILPLL provides at least 10 dB more reduction than SILPLL until 20 kHz offset. For comparison, phase noise of DSIL with 3km and 5km delays is also shown in Fig. 4 with magenta curve. We can see that phase noise of DSILPLL is 29 dB lower than that of DSIL at 300 Hz offset, which demonstrates the advantage of DSILPLL over DSIL alone.

DSILPLL also demonstrated better side-mode suppression as opposed to single loop SILPLL. In Fig. 4, the blue and green rectangles represent the actual level of the first spurious signals of SILPLL and DSILPLL, respectively. For SILPLL, the first spurious signal appears at offset frequency of 186 kHz with a level of -40 dBc (blue rectangle), which is relatively high; but for DSILPLL, the actual first spurious signal appears at 75 kHz offset with a level of -62 dBc (green rectangle), corresponding to a suppression of 22 dB compared to that in SILPLL configuration.

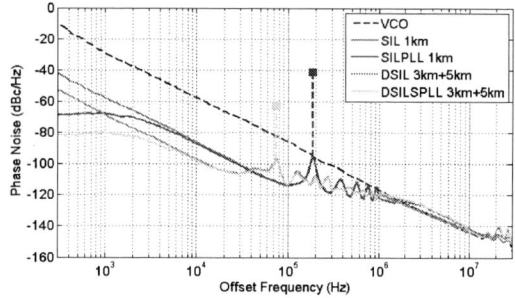

Fig. 4. Comparison of VCO phase noise employing different stabilization techniques. Black dashed: VCO free run; Red: SIL 1km; Blue: SILPLL 1km; Magenta: DSIL 3km+5km; Green: DSILPLL 3km+5km. SIL parameters: ρ = 0.003, ω_{3dB} = 60 MHz. SPLL parameters: Kd = 0.1V/rad, Ko = $2\pi \times 1$MHz/V

Measured phase noise of DSILPLL with various delay combinations are shown in Fig. 5. From the measured results, combination of 3 km and 5 km delays yields the lowest phase noise. Phase noise performance in the case of 1 km and 5 km delay is inferior to 3 km and 5 km delay combination even though the length of the longer delay is the same. This could be due to the side-modes of 1 km (every 200 kHz) and 5 km (every 40 kHz) being harmonically related thereby being less effective in side-mode suppression as opposed to the case of 3 km and 5 km delays, where the side-modes are non-harmonically related. In particular, 66.7 kHz for 3 km delay and 40 kHz for 5 km delay.

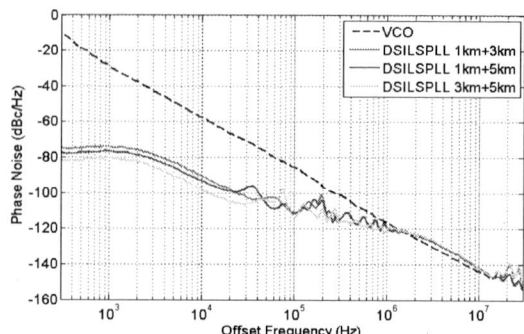

Fig. 5. Phase Noise of DSILPLL VCO with different length combinations. Black: VCO free run; Red: DSILSPLL 1km+3km; Blue: DSILSPLL 1km+5km; Green: DSILSPLL 3km+5km. SIL parameters: ρ = 0.003, ω_{3dB} = 60 MHz. SPLL parameters: Kd = 0.1V/rad, Ko = $2\pi \times 1$MHz/V

IV. DISCUSSIONS

Experimental performance evaluation of SILPLL oscillators using long optical delay lines based on patented techniques [10]-[11] is presented for the first time. From the measured results, it is advantageous by simultaneously combining SIL and SPLL techniques to achieve a cleaner spectrum in a wider offset range. In addition, dual loop configuration provides improved side-mode suppression as opposed to single loop.

REFERENCES

[1] H. P. Moyer et al., "A unified analytical model and experimental validations of injection-locking processes," *IEEE Trans. MTT.*, vol. 48, no. 4, pp. 493-499, Apr. 2000.

[2] C. McNeilage et al., "Review of feedback and feedforward noise reduction techniques," in *Proc. IEEE Freq. Control Symp.*, Pasadena, CA, May 1998, pp. 146 – 155.

[3] X.S. Zhou et al., "A new approach for a phase controlled self-oscillating mixer," *IEEE Trans. MTT.*, vol. 45, no. 2, pp. 196-204, Feb. 1997.

[4] D. Surzbecher et al., "Optically controlled oscillators for millimeter-wave phased-array antennas," *IEEE Trans. MTT.*, vol. 41, no. 6/7, pp. 998-1004, Jun/Jul 1993

[5] H. C. Chang, "Phase noise in self-injection locked oscillators – Theory and Experiment," *IEEE Trans. MTT.*, vol. 51, no. 9, pp. 1994 – 1999, Sep. 2003.

[6] L. Zhang et al., "Phase noise reduction and spurious suppression in microwave oscillators utilizing self-injection loops," in *Proc. Radio and Wireless Symp.*, Newport Beach, CA, USA, Jan. 19 – 23, 2014

[7] L. Zhang et al., "Analytical and experimental evaluation of SSB phase noise reduction in self-injection locked oscillators using optical delay loops," *IEEE Photonics J.*, vol. 5, no. 6, Dec. 2013

[8] G. Pillet et al., "Dual-frequency laser at 1.5μm for optical distribution and generation of high-purity microwave signals," *J. of Lightw. Technol.*, vol. 26, no. 15, pp. 2764-2773, Aug. 2008.

[9] L. Zhang et al., "Comparison of optical self-phase locked loop techniques for frequency stabilization of osicllators," (accepted for pulication on IEEE Photonics Journal)

[10] A. Poddar et al., Self Injection Locked Phase Locked Loop Optoelectronic Oscillator", US Patent application No. 61/746919.

[11] A. Poddar et al., "Integrated production of self injection locked self phase loop locked Opto-electronic Oscillators", US Patent app. 13/760767.

Gold Nanorod Array Structured Silicon Nitride Films for Reliable RF MEMS Capacitive Switches

L. Michalas[1], S. Xavier[2], M. Koutsoureli[1], O. El Jouaidis[2], S. Bansropun[2], G. Papaioannou[1], A. Ziaei[2]

[1] Solid State Physics Dpt., University of Athens, Athens, Greece 15784
Tel.: +302107276817, Fax: +302107276711, Email. lmichal@phys.uoa.gr
[2] Thales Research & Technology, F-91767 Palaiseau cedex, France

Abstract — The electrical properties of gold rods nanostructured silicon nitride are investigated. The paper aims to determine the advantages of the nanostructured material over conventional dielectrics that will mitigate the dielectric charging and provide a potential candidate for insulating films in MEMS capacitive switches. Different nanorod diameters and densities were grown. A model was implemented to describe both the DC and low frequency electrical properties. Finally, the device performance up to 40GHz was assessed.

Index Terms - Silicon nitride, RF MEMS, Nanostructured dielectrics, Dielectric charging, Reliability, Gold nanorods.

I. INTRODUCTION

Capacitive RF MEMS switches are one of the most promising applications in microelectromechanical systems (MEMS). In spite of this they have not yet reached the perfection to be characterized as "component of the shelf" due to reliability problems the most important of which is still the charging of the dielectric [1]. On the way to mitigate the problem of dielectric charging, several solutions have been proposed, such as the removal of dielectric film [2], the use of side actuation pads and exposing the CPW capacitor to RF signal only [3], the use insulator-insulator contact [4], etc.

The first attempt to increase the reliability of MEMS capacitive switches by introducing a nanostructured dielectric material, particularly silicon nitride with carbon nanotubes, was presented in [5]. At this point it is essential to point out that the carbon nanotubes were encapsulated between a 70 nm thick SiN layer that was deposited on the CPW and a second 180 nm thick SiN layer. The resulting material exhibited a larger "Figure of Merit" [5] that arose from the dielectric conductivity due to CNT density which was increasing and gradually approaching the percolation threshold [6]. The advantage of naturally nanostructured dielectric materials such as nanocrystalline diamond has been presented in [7,8]. Finally the importance of both naturally and engineered nanostructured dielectrics has been discussed in [6].

In view of these the aim of the present work is to fabricate a nanostructured dielectric film where highly ordered gold nanorods have been grown. Both the diameter and the density of nanorods are precisely controlled. The electrical properties of the resulting

material have been compared with those of a reference sample. The results have shown that the RF performance, assessed up to 40GHz, is not affected by the presence of nanorods.

II. FABRICATION AND ELECTRICAL ASSESSMENT

The nanostructured dielectric material was fabricated in practically three steps. A 100 nm SiN layer was deposited with PECVD method on bottom contact (CPW line for frozen MEMS), which was deposited on SiO_2/high resistivity Si substrate. Holes with diameter of 150 nm and 500 nm were opened in the SiN film and 100 nm Au nanorods were grown directly on the bottom contact. Finally, the nanostructured dielectric was covered with 100 nm PECVD SiN. The nanorods spacing was 1, 2.5, 5 and 10 μm, summarized in Table I and shown in Fig.1.

TABLE I. DIELECTRIC FILM STRUCTURE

	Density 1	Density 2	Density 3	Density 4
Diameter	500nm or 150nm	500nm or 150nm	500nm or 150nm	500nm or 150nm
Step	1μm	2.5μm	5μm	10μm

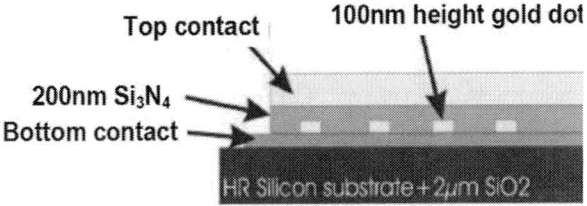

Fig.1 MIM and dielectric film structure

The layout of the Metal-Insulator-Metal (MIM) capacitors and frozen MEMS, the latter fabricated with different contact areas, with the above dielectric film as

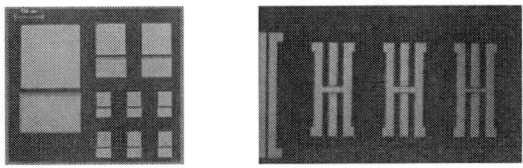

Fig.2 Layout of (a) MIM capacitors and (b) frozen MEMS capacitive switch

well as with the reference one is presented in Fig.2. Finally, the MIM capacitors were assessed by obtaining the current-voltage characteristics and charging current transient (CCT) with a Keithley 6517A electrometer. The device capacitance-voltage characteristics were obtained with the aid of a 1MHz Boonton 72B capacitance meter. Finally, the RF isolation of the frozen MEMS was assessed up to 40GHz.

III. DIELECTRIC FILM STRUCTURE AND ELECTRICAL PROPERTIES

In silicon nitride, the charge transport is achieved through percolation [6]. Other mechanisms such as Frenkel-Poole, hopping etc can also be identified and contribute if the temperature dependence of electrical properties is monitored [9]. The control of dielectric film deposition with the additional inclusion of nanoparticles may further improve the dielectric film properties regarding the charge injection and charge removal procedures. Further improvement can be achieved by introducing additional conductive paths such conductive nanorods that are normal to bottom electrode. These locally increase the electric field, during both the charging and discharging increasing the charge collection over a finite area at the surface of the dielectric film.

In the present work it was chosen to include gold nanorods with different diameters and densities in order to determine the optimum structure for reliable MEMS capacitors. The study of the electrical characteristics adopted a classical physics model for the DC and low frequency electrical properties of the MIM capacitors. In order to derive the model we assumed that the nanorods form cells with dimension D in the center of which there is only one nanorod of diameter d. The distance between the top and the bottom electrode is H and from the top of the nanorod to the top electrode is h (Fig.3)

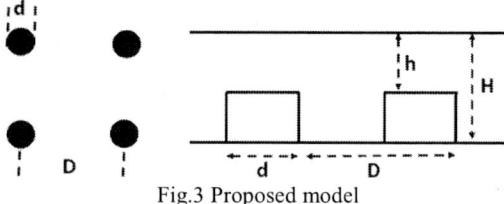

Fig.3 Proposed model

If the electric field fringing is neglected then the capacitance per unit cell will be given by the sum of the capacitance of the top area of the nanorod (C_{NR}) and the capacitance of the bottom electrode (C_{BE}), with respect to the top electrode. Thus the measured capacitance per unit area ($C_{MIM-meas}$) will be equal to the capacitance per unit cell and it will be given by:

$$C_{MIM-meas} = \frac{\varepsilon}{H}\left[1 + \frac{\pi d^2}{4D^2}\cdot\left(\frac{H}{h}-1\right)\right] \qquad (1)$$

Furthermore, the measured current will arise from the sum of the current from the top of the nanorod (J_{NR}) and the bottom electrode (J_{BE}). According to the proposed model, if the fringing electric field is low then the plot of measured capacitance for a series of samples with constant nanorod diameter vs inverse cell area will be linear: $C_{MIM-meas} \propto \dfrac{d^2}{D^2}$. Regarding the measured average current density it can be written as:

$$J_{MIM-meas} = J_{BE} + \left(J_{NR} - J_{BE}\right)\cdot\frac{\pi}{4}\cdot\frac{d^2}{D^2} \qquad (2)$$

that suggests a linear relation to inverse cell area.

The plot of capacitance and current density corresponding to unit cell vs inverse cell area are presented in Fig.4. The fact that the line intercepts the origin of the axis indicated that the current density J_{BE} is negligible and therefore the charging and discharging currents will be determined by the nanorods density and diameter. Here it must be pointed out that the fringing

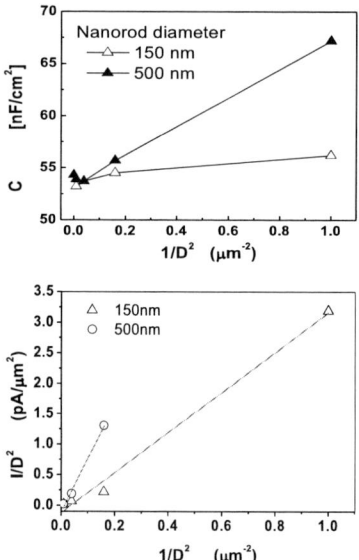

Fig 4 Unit cell capacitance and current dependence on cell dimension

electric field, hence fringing capacitance, should be manifested strongly in small cells.

In spite of the above conclusions the contribution of the fringing field to both charging and discharging was tested by comparing the current density J_{NR} for the 150 nm and

978-1-4799-8198-4/15 $31.00 © 2015 IEEE

500 nm rods. The comparison of the calculated values indicates that the fringing field of the thin nanorods increases the diameter of 150 nm to an effective value of about 310 nm. Moreover, taking into account that the ratio of the top areas of the nanorods is $A_{500nm}/A_{150nm} \cong 11$ we are led to the conclusion that the electric field fringing becomes important for thin nanorods and therefore for applications employing carbon nanotubes. The diameter difference is the experimentally extracted threshold. Thus it allows the prediction of minimum required distance between the nanorods in order to achieve non negligible fringing field overlapping and therefore succeed charge collection without increasing significantly the pull-in leakage current. This result does not revoke the proposed model because in the present work the smallest cell had dimensions that did not allow such interference.

The dependence of isolation on frequency, up to 40GHz for frozen MEMS with contact area of 2×10^{-4} cm^2 and nanorod densities 1, 3 for 500 nm diameter and 4 for 150 nm diameter (see Table I) including the reference sample, which does not contain nanorods, is shown in Fig.5. The obtained plots clearly show that the presence of nanorods does not affect the RF performance and on the other hand increase the device reliability.

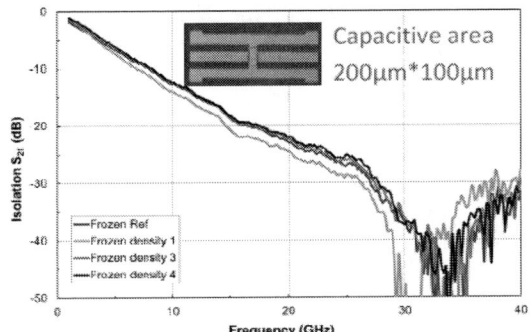

Fig.5 Frequency dependence of isolation S_{21} of a frozen MEMS

IV. CONCLUSION

The effect of the introduction of the conductive (Au) nanorods on silicon nitride electrical properties has been investigated. It has been shown that if the nanorods distribution and dimensions is precisely controlled, a simple model can be used, which is based on classical physics and it can model efficiently the DC and low frequency properties of MIM capacitors and therefore the down-state MEMS. The model does not take into account the fringing field effects and showed excellent agreement to the experimental results from the fabricated structures. The model allows the determination of the threshold distance between the nanorods below which the fringing field overlapping becomes significant. The threshold

distance allows the design of nanostructured dielectrics that achieve optimum charge collection and in the same time maintain, during pull-in, the leakage current low enough so that to avoid the device burnout. It is interesting to notice that the exploitation of experimental data with the proposed model can be extended to even thinner nanorods if the quantum effects at the interface with the dielectric material still remain low enough.

ACKNOWLEDGEMENT

The authors wish to acknowledge the support from project: "Nanostructured materials and RF-MEMS RFIC/MMIC technologies for highly adaptive and reliable RF systems" NANOTEC under GA 288531

REFERENCES

[1] G. M. Rebeiz, "RF MEMS *Theory, Design and Technology*, Haboken, New Jersey: J. Willey and Sons, 2003

[2] D. Mardivirin, A. Pothier, A. Crunteanu, B. Vialle, P. Blondy, "Charging in Dielectricless Capacitive RF-MEMS Switches", IEEE Trans. on Mic. Theory and Tech., vol. 57, pp. 231-236, 2009

[3] Kim Sangchae, J. McKillop, A. Morris, "Characterization of dielectric charging and reliability in capacitive RF MEMS switches" in *Proc. 2013 IEEE Int. Reliab. Physics Symposium (IRPS)*, Monterey, CA, pp. 6.B4.1-6B.4.5, 2013

[4] D. Molinero, S. Cunningham, D. DeReus, A. Morris, "Dielectric charging characterization in MEMS switches with insulator-insulator contact", in *Proc. 2014 IEEE International Reliability Physics Symposium (IRPS)*, Waikoloa, HI, 2014, pp. 5C.3.1 - 5C.3.4

[5] C. Bordas, K. Grenier, D. Dubuc, E. Flahaut, S. Pacchini, M. Paillard, J. –L. Cazaux, "Carbon Nanotube Based Dielectric for Enhanced RF MEMS Reliability", in *Proc. 2007. IEEE/MTT-S International Microwave Symposium*, Honolulu, HI, 2007, pp. 375-378.

[6] G Papaioannou, L. Michalas, M. Koutsoureli, S. Bansropun, A. Gantis and A. Ziaei, "Charging Mechanisms in Nanostructured Dielectrics for MEMS Capacitive Switches", in *Proc. of 2014 IEEE 14th Topical Meeting on Silicon Monolithic Integrated Circuits in Rf Systems (SiRF)*, Newport Beach, CA, 2014, pp. 98-100

[7] C. Goldsmith et al., "Charging characteristics of ultra-nano-crystalline diamond in RF MEMS capacitive switches", in *Proc. 2010 IEEE MTT-S International Microwave Symposium Digest (MTT)*, Anaheim, CA, 2010, pp. 1246-1249

[8] L. Michalas, M. Koutsoureli, S. Saada, C. Mer–Calfati, A. Leuliet, P. Martins, S. Bansropun, G. Papaioannou, P. Bergonzo and A. Ziaei, "Electrical assessment of diamond MIM capacitors and modeling of MEMS capacitive switches discharging", Journal of Micromechanics and Microengineering, vol.24, art.115017, 2014

[9] L. Michalas, et al., "Electrical characterization of undoped diamond films for RF MEMS application", in *Proc. 2013 IEEE International Reliability Physics Symposium (IRPS)*, Monterey CA, 2013, pp. 6B.3.1-6B.3.7

Miniaturized 60 GHz Triangular CMOS Antenna-on-Chip using Asymmetric Artificial Magnetic Conductor

Adel Barakat [1], Ahmed Allam [1], Hala Elsadek [2], Adel B. Abdel-Rahman [1], S. Muhammad Hanif [3], and Ramesh K. Pokharel [3]

[1] Egypt-Japan University of Science and Technology, New BorgAlarab 21934, Egypt
[2] Electronics Research Institute, Giza 21622, Egypt
[3] Kyushu University, Fukuoka 819-0395, Japan

Abstract —This paper presents a miniaturized triangular Antenna-on-Chip (AoC) designed and fabricated on a 0.18 μm CMOS process using asymmetric rectangular artificial magnetic conductor (R-AMC). An AMC acts as a shield plane between the AoC and the lossy CMOS substrate. AoC using asymmetric R-AMC presents a smaller overall area than that of the previous reported AoC with symmetric AMCs. The triangular AoC area including the asymmetric AMC cells is only 0.81mm^2 with a simulated gain of -0.2 dBi at 60 GHz. Measurements confirm the wide impedance bandwidth of the AoC.

Index Terms —60 GHz, TSMC, CMOS, Antenna-on-Chip (AoC), Artificial Magnetic Conductor (AMC).

I. INTRODUCTION

The 60 GHz band is known by its wide bandwidth of 7 GHz from 57 GHz to 64 GHz. This wide bandwidth will allow for high data rate applications at the level of Gbps suppressing the current wireless technologies. A variety of applications such as uncompressed high definition video streaming, mobile distributed computing, wireless gaming, internet access, fast large file transfer, Wireless Personal Area Networks (WPANs), etc. fall into the 60 GHz communications interest. Communications at 60 GHz have several advantages such as the possibility of frequency reuse over small distances due to the high attenuation by atmospheric oxygen of 10 to 15 dB/km and the prospect of antenna miniaturization where the corresponding wavelength in free space is only 5 mm at 60 GHz [1]-[6].

Antenna-on-Chip (AoC) integration with other radio frequency and digital circuits is a prime target for a complete System-on-Chip. However, Complementary-Metal-Oxide-Semiconductor (CMOS) AoC integration suffers from two main disadvantages. First, AoC suffers from losses due to low resistivity CMOS substrate (σ=10S/m) which leads to reduction of the AoC efficiency [2]-[6]. Second, AoC occupies a large area that cannot be reused because of the strong field below the AoC. In order to increase AoC efficiency, Artificial Magnetic Conductor is used to electromagnetically shield the AoC from the lossy CMOS substrate. This is possible because of the AMC reflection characteristics. AMC has a reflection magnitude of +1 and a reflection phase of zero at its center

frequency and varies between ±180°. The useful bandwidth of the AMC is between ±90° of reflection phase at which constructive interference occurs between incident wave and reflected wave on AMC surface [1]-[6]. Although AMC is used to enhance efficiency, it results in an increase of the overall design area such as 2.4 mm^2 patch antenna with snowflake AMC presented in [5]. In [4], an optimization methodology is proposed to increase AoC efficiency using AMC while maintaining small overall area. This is possible by increasing number of AMC cells in the current flow direction and limiting the number of AMC cells in the perpendicular direction. Using this technique, triangular AoC is designed with symmetric rectangular (R)-AMC in [3] with overall area of 1.51 mm^2.

In this paper, a small size triangular AoC using asymmetric R-AMC is designed and fabricated on TSMC 0.18 μm CMOS technology. The design was optimized using High Frequency Structure Simulator, HFSS® and finally, fabricated using the 0.18 μm CMOS technology and measured. The paper is arranged as follows: section II presents the asymmetric AMC unit cell designs. Section III describes the design of the AoC with the asymmetric AMC. Furthermore, section IV details the fabrication and measurements. Finally, the paper concludes with main results in section V.

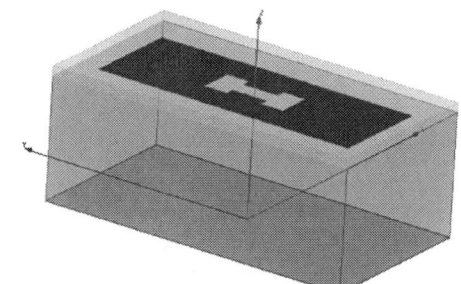

Fig. 1.   3D view of asymmetric R-AMC cell

II. ASYMMETRIC AMC UNIT CELL DESIGN

An asymmetric AMC unit cell three dimensional (3D) view is shown in Fig. 1. An asymmetric R-AMC is proposed to allow for further miniaturization of the overall design dimensions by limiting the AMC unit cell size in the direction at which the AMC property is not required. The Phase Response (PR) of this unit cell is shown in Fig. 2.The asymmetric R-AMC unit cell is limited in size in the X-direction and only show AMC properties for the Y-polarized signals at the bandwidth of interest. While for X-Polarized signals it shows AMC properties at higher frequencies.

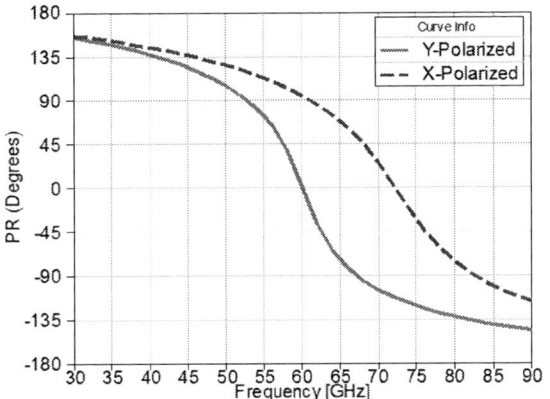

Fig. 2.   Asymmetric R-AMC cell phase response

III. Triangular AoC Design

Three dimensional view of the triangular AoC with 4×6 cells of asymmetric R-AMC is shown in Fig. 3. The number of cells is optimized following the methodology presented in [4]. The top metal layer (M6) is used for the triangular AoC to ensure low conduction loss because M6 has the largest copper thickness when compared to the other metal layers available on the TSMC six metal layer stack. Bottom metal layer (M1) is used for the design of the asymmetric AMC cells. The overall design size including the asymmetric AMC cells is 0.81 mm2 which is possible because of the limited size of asymmetric AMC cells in the X-direction, besides the use of technique presented in [4].

Fig. 3.   3D view of triangular AoC over 4 × 6 cells of asymmetric R-AMC

Simulated gain and efficiency of the triangular AoC using asymmetric AMC are shown in Fig. 4. A peak

simulated gain of -0.2 dBi and a simulated efficiency of 44% are possible at 60 GHz.

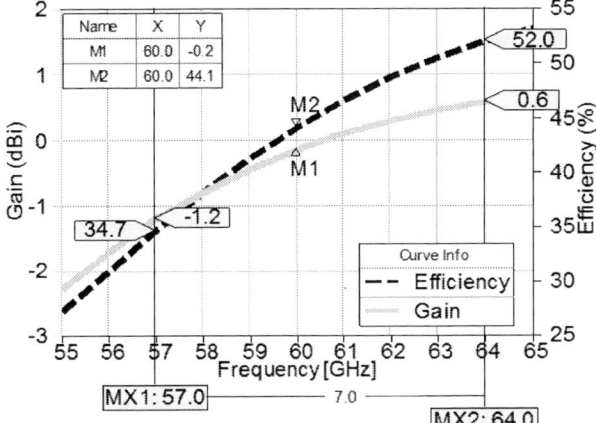

Fig. 4.   Simulated gain and efficiency of triangular AoC over 4 × 6 cells of asymmetric R-AMC

In order to compare the performance of this AoC with previously reported AoCs, Figure-of-Merit (*FoM*) presented in [5] is used. This *FoM* can be computed as:

$$FoM = \frac{Absolute\ Gain}{Area} \times 100 \qquad (1)$$

As shown in table I, the proposed triangular AoC using asymmetric AMC has the highest *FoM* and it can be considered the best design.

TABLE I
COMPARISON BETWEEN DIFFERENT AoC IN LITERATURE WITH THIS WORK

Design	Gain(dBi) @ 60 GHz	Area (mm^2)	*FoM*
This work	-0.2	0.81	118
[2]	0.8	1.51	80
[3]	0	1.39	72
[4]	-2.4	2.4	24

IV. Fabrication and Measurements

The triangular AoC using asymmetric AMC is fabricated using TSMC 0.18 μm CMOS technology. Fabricated AoC chip photo is shown in Fig. 5. Measurements are done using Agilent E8361C PNA and manual probe station. S11 measurements are shown in Fig. 6.

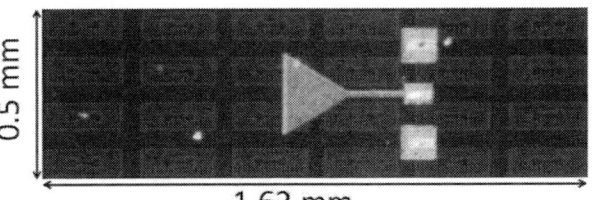

Fig. 5.   Chip photo of triangular AoC over 4 × 6 cells of asymmetric R-AMC

The measured S11 shows very wide bandwidth from less than 40 GHz to 65 GHz. This wide bandwidth means lower quality factor. The reason for the low quality factor may be the additional substrate losses as the design is simulated on substrate with a small area of 0.81mm^2 and fabricated on 25mm^2 substrate. While, the AoC has a simulated bandwidth from 50 GHz to 70 GHz. The shift in resonance may be because of the parasitic capacitances due to the surrounding circuits' metal layers fabricated together with the AoC and not considered in simulations.

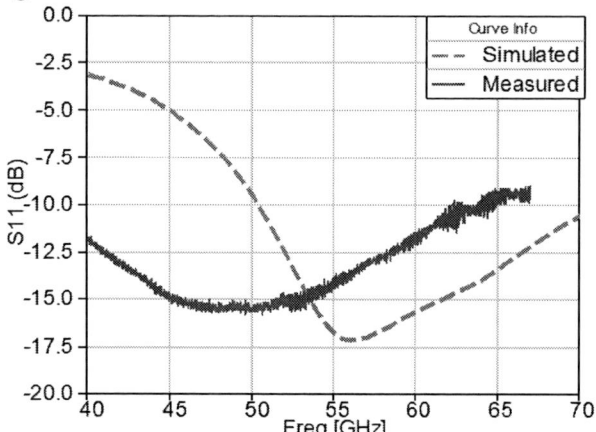

Fig. 6.   Simulated and measured S11 of triangular AoC over 4 × 6 cells of asymmetric R-AMC

V. CONCLUSION

A triangular AoC is presented for 60 GHz band. The triangular AoC is designed using a standard 0.18μm CMOS process and optimized with asymmetric R-AMC. Asymmetric R-AMC shows enhanced radiation characteristics when used as a shield. It contributes in design miniaturization. The fabricated AoC overall size including R-AMC is 0.5 mm by 1.62 mm and has a simulated gain and efficiency of -0.2 dBi and 44%, respectively at 60 GHz. The measured return loss confirms the operation at 60 GHz.

ACKNOWLEDGEMENT

This work is supported by Egyptian ministry of higher education, mission department.

Special thanks to Prof. H. Kanaya of Kyushu University for their support in fabrication and measurements.

REFERENCES

[1] P. Smulders, "Exploring the 60 GHz band for local wireless multimedia access: Prospects and future directions," *IEEE Comm. Magazine*, vol. 40, no. 1, pp. 140 - 147, 2002.

[2] H. M. Cheema, A. Shamim, "The last barrier: on-chip antennas," *IEEE Microwave Magazine.*, vol.14, no.1, pp.79 – 91, Jan.-Feb. 2013

[3] A. Barakat, A. Allam, R.K. Pokharel,H. Elsadek,M. El-Sayed, andK. Yoshida, "Compact size high gain AoC using rectangular AMC in CMOS for 60 GHz millimeter wave applications," in *Microwave Symposium Digest (IMS), 2013 IEEE MTT-S International*, 2013,pp.1-3

[4] A. Barakat, A. Allam, R. K. Pokharel, H. Elsadek, M. El-Sayed, and K. Yoshida, "Performance Optimization of a 60 GHz Antenna-on-Chip over an Artificial Magnetic Conductor," *in Electronics, Communications and Computers (JEC-ECC), 2012 Japan-Egypt Conference on*, 2012, pp.118-121

[5] H. Chu, Y. X. Guo, F. Lin, and X. Q. Shi, "Wideband 60GHz On-Chip Antenna with an Artificial Magnetic Conductor," *in Radio-Frequency Integration Technology (RFIT), 2009 IEEE International Symposium on*, 2009, pp.307-310

[6] A. Barakat, A. Allam, R. K. Pokharel, H. Elsadek, M. El-Sayed and K. Yoshida, "60 GHz triangular monopole Antenna-on-Chip over an Artificial Magnetic Conductor," *Antennas and Propagation (EUCAP), 2012 6th European Conference on* , 2012, pp. 972-976

A

Abdel-Rahman, Adel B. 92
Abou-Khalil, Michel 30
Alazemi, A. ....(NA)
Allam, Ahmed 92
Aufinger, Klaus 4
Avser, B. ....(NA)

B

Bansropun, S. 89
Barakat, Adel 92
Belot, Didier 58
Bendixen, J. 27
Ben Yishay, Roee 15
Botula, Alan 30

C

Cai, Zhiyong 83
Carroll, M. 27
Chang, Da-Chiang 11
Chang, Tzu-Hsuan 83
Chang, Yin-Cheng 11
Chao, Tzu-Yuan 21
Chen, Wei-Cheng 21
Cheng, Chih-Chieh 37
Cho, Seong Jun 18, 76
Cho, Y. ....(NA)
Chou, Min-Chih 11
Coccetti, Fabio 70
Colestock, P. ....(NA)
Costa, J. 27

D

Daryoush, Afshin S. 86
Deltimple, Nathalie 58
De Paolis, Rosa 70
Desbonnets, Eric 33

E

Elad, Danny 15
El-Gabaly, Ahmed M. 49
El Jouaidis, O. 89
Ellis-Monaghan, John 30
Elsadek, Hala 92

F

Fung, Helen 73

G

Gambino, Jeffrey 30
Genc, Alper 37
Gong, Shaoqin 83
Granger-Jones, M. 27
Gross, Jeff 30
Guegan, Guillaume 70
Gurbuz, O. .... (NA)

H

Hanif, S. Muhammad 92
Hasenaecker, Gregor 4
He, Zhong-Xiang 30
Helmi, Sultan R. 52
Hsu, Shawn S. H. 11

I

Ishikuro, Hiroki 24
Iversen, C. 27

J

Jaffe, Mark 30
Joseph, Alvin 30
Jou, Alice Yi-Szu 67

K

Kahmen, Gerhard 40
Kang, Dongmin 18, 76
Kasper, Erich 1
Kawai, Seitaro 43, 46
Kerherve, Eric 58
Kerr, D. 27
Kim, Hong Yi 18, 76
Kim, Samuel 73
Kimura, Kento 80
Kitazawa, Naoki 24
Ko, C.H. .... (NA)
Kohira, Kaoru 24
Koutsoureli, M. 89
Kuo, Chien-Nan 21

SiRF 2015 Brief Author Index

L

Lai, Chih-Wei 21
Larie, Aurélien 58
Larson, L. 61
Le, Dustin 73
Lee, Chae Jun 18, 76
Lee, Hae Jin 18, 76
Lee, Joong Geun 18, 76
Li, Chun-Hsing 21
Li, Jun 7
Lim, Kimsrun 46
Liu, Chen 67

M

Ma, Zhenqiang 83
Maglione, Mario 70
Martineau, Baudouin 58
Mason, P. 27
Matsuzawa, Akira 43, 46, 80
Michalas, L. 89
Mohammadi, Saeed 52, 67
Moon, Jeong-Sun 73
Moret, Boris 58
Morris III, Arthur S. 63
Musch, Thomas 4

N

O

Oh, Inn-Yeal 18, 76
Oh, Thomas 73
Okada, Kenichi 43, 46, 80

P

Papaioannou, G. 89
Park, Chul Soon 18, 76
Payan, Sandrine 70
Phelps, Richard 30
Poddar, Ajay K. 86
Pohl, Nils 4
Pokharel, Ramesh K. 92

Q

R

Ranta, Tero 37
Raskin, Jean-Pierre 33
Rebeiz, G.M. .... (NA)
Rein, Hans-Martin 4
Rhee, Woogeun 7
Rohde, Ulrich L. 86
Rotella, Francis 37

SiRF 2015 Brief Author Index

S

Saavedra, Carlos E. 49
Sabo, Ronald 83
Schmitz, Adele 73
Schumacher, Hermann 40
Sekar, Vikram 37
Seo, Hwa-Chang 73
Seo, Jung-Hun 83
Seo, Yuuki 46
Shan, Hengying 52
Shank, Steven 30
Shen, Yun-Chun 11
Shumaker, Evgeny 15
Slinkman, James 30
Son, Kyung-Ah 73
Song, In Sang 18, 76
Song, Yonghoon 55
Spears, E. 27
Sun, Yuanfeng 7

T

Tokgoz, Korkut Kaan 43, 46

U

V

W

Wang, Huei 14
Wang, Ping-Yi 11
Wang, Yuanxun Ethan 55
Wang, Zhihua 7
Whatley, Richard 37
Wolf, Randy 30
Wu, Te-Lin 11

X

Xavier, S. 89
Xu, Ni 7

Y

Yan, Tzu-Chao 21
Yang, Baohua 73
Yoon, Chong Hyun 76

Z

Zeng, Chang 37
Zhang, Li 86
Zhang, Wogong 1
Zhu, Rui 55
Ziaei, A. 89

9781479981984